# CHE 111L

**THIRD EDITION**

# GENERAL CHEMISTRY
## LAB I


JEROME MAY
DEPARTMENT OF CHEMISTRY
EASTERN KENTUCKY UNIVERSITY

*General Chemistry Lab I*

CHE 111L
Jerome May | Third Edition
Department of Chemistry
Eastern Kentucky University



Printed in the United States of America
10 9 8 7 6 5 4 3
ISBN: 978-1-61740-750-5

Van-Griner Learning
Cincinnati, Ohio
www.van-griner.com

President: Dreis Van Landuyt
Project Manager: Janelle Lange
Customer Care Lead: Lauren Houseworth

May 750-5 Su19
312612-321649
Copyright © 2020

# Table of Contents

# The 20-Question Pre-Lab Assignments and Pre-Lab Videos



## 20-Question Pre-Lab Assignments

For the vast majority of the CHE 111L experiments and lab activities, there is a YouTube video (approximately 30 minutes long) for students to watch and a set of 20 questions whose answers are obtained by watching the video.

The purpose of the pre-lab video is to introduce CHE 111L students to the science behind the experiments before you attend lab so that you can be better prepared to fully understand the physical and chemical principles that we are investigating by performing this specific experiment.

The purpose of the 20-Question Pre-Lab Assignment is to help you identify the aspects of the information presented in the video that are the most important.

The following is a suggestion as to how I would approach this assignment if I were a CHE 111L student:

1. Read through the questions on the pre-lab assignment to see what I should be watching for.

2. Watch the video all the way through from start to finish uninterrupted.

3. Try to answer those questions that I could based on that first viewing.

4. Watch the video in "stop-and-go" fashion to fill in the answer gaps.

Naturally, the preceding is only a suggestion, and you are encouraged to complete the assignment in the manner that best suits you.

There is a list of links to all of the videos under the **Course Documents** content area, *and* each weekly **Announcement** also contains the link for that week's video. Further, the links are included with each experimental description under the **Experiments** content area.

Finally (and in summary), the 20-Question Pre-Lab Assignment will count as 20% of each lab report grade. (Yes, fully 20 points of each 100 point lab report form will be the 20-Question Pre-Lab Assignment for that week.)

You should go to the 20-Question Pre-Lab Assignment for each lab activity or experiment, watch the pre-lab video on YouTube, write your answers legibly, and bring your answers with you to lab.

Your copy of the 20-Question Pre-Lab Assignment *must* be handed to your TA as you walk in the door/settle into your lab station, or it will not be accepted. (Partially completed 20-Question Pre-Lab Assignments will be accepted for partial credit; but they, too, must be handed to your TA before lab begins.)

*No late 20-Question Pre-Lab Assignments will be accepted—no exemptions and no exceptions.*

# Pre-Lab Videos

### Activity 1: Logger Pro Graphing

*https://www.youtube.com/watch?v=gVlG0GwUr0Q*

### Activity 2: IUPAC Inorganic Nomenclature

*https://www.youtube.com/watch?v=vgqsqD-KVil*

### Experiment 1: Measurements and Density

*https://www.youtube.com/watch?v=7iIw8YyY5Sw*

### Experiment 2: The Determination of the Chemical Formula of a Hydrated Salt

*https://www.youtube.com/watch?v=ChcGo5al6uk*

### Experiment 3: Solution Conductivity: Electrolytes, Dissociation, and Ionization

*https://www.youtube.com/watch?v=xc3lRRNmm4c*

### Experiment 4: Chemical Reactions in Aqueous Solutions

*https://www.youtube.com/watch?v=Elislow4Xpg*

### Experiment 5: Solution Stoichiometry

*https://www.youtube.com/watch?v=_fcPLHmE_pE*

### Experiment 6: Oxidation–Reduction Reactions

*https://www.youtube.com/watch?v=PEsK6O64Oe0*

### Experiment 7: Gas Laws

*https://www.youtube.com/watch?v=QKMwMXf5j5E*

### Experiment 8: Gas Stoichiometry

*https://www.youtube.com/watch?v=QKMwMXf5j5E*

### Experiment 9: Calorimetry: The Determination of the Specific Heat of a Metal

*https://www.youtube.com/watch?v=4p2OBOba5dc*

### Experiment 10: Atomic Structure, Periodicity, Waves, and Emission Spectroscopy

*https://www.youtube.com/watch?v=q5ceskeUyYA*

### Experiment 11: Drawing LEDDs of Molecules

*https://www.youtube.com/watch?v=PRTzZxvAGEA*

# Logger Pro Graphing



## Objectives

At the completion of the activity, you should be able to

- produce a graph that is properly labeled and scaled using the Logger Pro program;

- recognize a direct and an inverse graphical relationship;

- determine the equation for the straight-line relating two variables and give the meaning of the slope and intercept for a straight-line; and

- recognize an exponential relationship between two variables.

# 20-QUESTION PRE-LAB ASSIGNMENT

CRN:_________________________________  Your Name: _______________________________

Date Submitted: ______________________  TA's Name: _______________________________

*Please write your answers legibly in the space provided. Some answers will actually consist of three or four pieces of information that are clearly written on the board during the conversation **or** clearly stated verbally during the presentation. You must include all aspects of the answer to receive full credit. These answers are due at the beginning of the lab meeting.*

1. What is the Chemistry Department's preferred spreadsheet program?

2. What are two names for the vertical axis?

3. What are two names for the horizontal axis?

4. How **must** the title of a scientific graph be written?

5. For sporting events listed as team one vs. team two, which team is the home team?

6. How is the $x$-axis variable "defined"?

7. What is the classic example of an $x$-axis variable quantity?

8. How is the $y$-axis variable defined?

**9.** What are three typical shapes possible for an $x, y$-plot?

**10.** For which of the above three shapes do the plotted data approach a limiting value?

**11.** What is the equation for a linear plot? Identify all the terms in the equation.

**12.** How do we calculate the slope, $m$, of the linear plot?

**13.** What is the equation for a parabolic plot?

**14.** What is the easiest way to obtain the slope of a parabolic plot (even though we do not use it in CHE 111L)?

**15.** How do we linearize parabolic data?

**16.** Draw a sketch below to show how a parabolic plot becomes a linear plot.

**17.** What can Logger Pro do for us with a hyperbolic (and/or a parabolic) plot of data?

**18.** What is our general (fractional) formula to treat proportions? Show how we did this for a group of 50 human males plus 50 human females.

**19.** Show the proportion "calculation" for a group of 25 people wearing silver plus 50 people wearing maroon.

**20.** Should we include the dollar sign ($) when we plot salary data in Logger Pro?

# BACKGROUND

In this exercise, you will be learning how to make graphs using Logger Pro software. This software is the preferred data analysis software for General Chemistry at EKU and is on the computer in the Chemistry Computer Lab (NSB 4104) and Tutoring Center (NSB 5103), as well as in the laboratory rooms (NSB 4118 and NSB 4122).

## Basic Information about Graphs

Making graphs correctly is an important skill for any science and for many other fields of study. A graph is a way to show the relationship between two data values, which we usually call variables. There are many different types of graphs, such as line, bar, pie, and scatter. In our labs, we will most often be making *scatter plots.*

In a scatter plot, the variable that changes steadily and incrementally—and *sometimes* beyond human control (the *independent variable*) is plotted on the *x*-axis. The variable that responds to changes in the independent variable is called the *dependent variable.* It is plotted on the *y*-axis. For example, the pay you get from working at a job is dependent on the number of hours you work. So, the number of hours is the independent variable, and your pay is the dependent variable. These two variables are *directly* related: as one increases, the other increases. Other variables are *inversely* related. An example of that would be the pressure and volume of a gas. If you increase the pressure on a container of gas, the volume will decrease.

In the first and second parts of this lab, you will use the program called **Logger Pro** to plot a set of data you are given.

# PROCEDURE

## Logger Pro Quick Start Guide

1. Open Logger Pro version 3.10 (version 3.9 on some computers) by double clicking on the icon on the desktop. This will open a screen that has a blank data table to the far left and a 2D set of Cartesian coordinate axes for the graph on the right.

2. Enter the data points in the table. They will automatically be plotted on the graph.

3. You should always label the $x$ and $y$-axes of a graph. To do this in Logger Pro, double click on the "X" at the top of the $x$-axis data. A new box will open called "manual column options." Select the "column definition" tab. Enter the name of the quantity you are plotting. Below the "name" box is a place to add units. Always include **units** with the axis labels.

4. While you are in this box, click on the "options" tab. You can select the size, style, and color of the points that are plotted on the graph.

5. Also in the "options" tab, you could specify the number of decimal places or significant figures you want to be displayed on the axis.

6. Repeat this for the $y$-axis labels.

7. For this experiment, you do not want the points to be connected with a line initially. If Logger Pro has already filled in a line, remove it by first clicking inside the graph area. From the top menu bar, select "options—graph options." You will get a new box with options. One of them is "appearance." If the "connect points" box is checked, remove the check. Then your graph will have only points.

8. In this same box, you can put a title on your graph. The graph title should ***always*** be expressed in the format of "<the $y$-axis variable> vs. <the $x$-axis variable>."

9. We want the plotted points on the graph to fill the majority of the space available for the graph. After your points are entered, click the autoscale icon from the top menu bar. (This is an "A.") Logger Pro will adjust your graph so that the region of the graph with the plotted points fills almost the entire screen.

10. To get a linear fit of the graph, click "analyze" from the top menu bar. Select "linear fit."

11. A box will appear on the plot that gives the slope and intercept of the line, along with a correlation label. A line will also be plotted on the graph.

12. To print, select "file—page setup." ***Be sure to select landscape mode!***

13. Then select "file—print graph." You will be able to enter a footer for the graph—you might want to use this to enter your name and the date.

# DATA SHEET

Name: _________________________________     Grade: _________________________________

Date Experiment Performed: _______________     Days Late: _____________________________

CRN of Lab Section: ____________________     Instructor's Initials: ___________________

| General Grading Items | 20 Points |
|---|---|
| | |
| All Safety Rules Were Followed | |
| Waste Was Properly Disposed of and Lab Area Was Cleaned | |
| Evaluation of Student Performance Overall (Student Was on Time, Followed Safety Rules, Performed the Lab Correctly and Within the Time Allowed, Etc.) | |
| 20-Question Pre-Lab Assignment | /20 |
| **Total** | /20 |

**CHE 111L student: Attach the copies of the graphs you made for this experiment to this report.**

# Part I: Directly Proportional and Linear (25 points)

1. Attach the graph you made of the pay vs. hours worked data. Be sure it is properly labeled and scaled.

2. The first graph plotted your pay for the number of hours worked. Is this a direct or inverse relationship?

3. What is the **general** equation for a straight-line on a graph?

4. What is the meaning of the letters *m* and *b* in the equation for a straight-line?

5. Define the slope of a line.

6. Write the equation for the straight-line on your graph.

7. What is the numerical value of the slope of the line you plotted?

8. What is the real-world meaning of the numerical value of the slope of this graph?

**9.** Define the term *y*-intercept.

**10.** What is the numerical value of the *y*-intercept on your graph?

**11.** What is the real-world interpretation of the *y*-intercept on this graph?

**12.** One use for graphs is the prediction of a parameter. Use the graph to predict your pay if you work 16.8 hours. ***Show how you worked this problem.***

## Part II: Parabolic or Hyperbolic? (definitely *non*-linear) (35 points)

**13.** Attach the graph you made of how the activity of a radioactive material changes over time. Be sure it is properly labeled and scaled.

**14.** This graph plots the activity of a radioactive material vs. the time that passes after it is first formed. Does this graph show that this a direct or inverse relationship?

**15.** Does this relationship appear to be a linear relationship?

**16.** What kind of relationship does it appear to be?

**17.** Examine the numbers in this set of data. Can you see any general trend in what happens to the activity every 15 seconds?

**18.** Predict the *approximate* activity of the radioactive material when 120 seconds has passed. **Show your reasoning for your answer.**

**19.** Sometimes it is possible to take data that are not linear and transform them in such a way that they become linear. Use your calculator and determine the **natural logarithm** (the "base e" logarithm, where e = 2.71828) of each value of the activity level.

Keep/record and **plot four digits** for each ln value.

    **a.** ln 5000 =                         **e.** ln 308 =

    **b.** ln 2490 =                        **f.** ln 161 =

    **c.** ln 1255 =                        **g.** ln 77 =

    **d.** ln 619 =                         **h.** ln 38 =

**20.** Now make a graph of the natural log of the activity values (on the *y*-axis) vs. the time (on the *x*-axis, in seconds). You can use Logger Pro or Excel to do this, or you can use a piece of graph paper and do the graph by hand. In either case, be sure that your graph is correctly labeled and scaled. Attach your labeled and scaled graph.

**21.** Is this graph linear?

**22.** Determine the slope and the *y*-intercept of the line, and write the equation for the line.

**23.** Use this equation to predict the *activity* (not the natural log of the activity) of the material at 120 seconds. **Show your work.**

**24.** Compare the answer you got in Question 23 with the answer you predicted for this time in Question 18.

# Report on Graphical Analysis (20 points)

**Fill in the table and write your answers to the questions in the spaces below.**

Consider two pure liquids A and B. We do not know the density of liquid B, but we do know the density of pure liquid A and the densities of the following mixtures of A and B.

| TABLE 1 | | | | |
|---|---|---|---|---|
| Moles of A* | Moles of B | Total Moles | Fraction of A | Density/g·mL$^{-1}$ |
| 1.00 | 4.00 | | | 1.33 |
| 2.00 | 3.00 | | | 1.29 |
| 3.00 | 2.00 | | | 1.20 |
| 4.00 | 1.00 | | | 1.11 |

*Pure liquid A: $d$ = 1.05 g/mL

**25.** Calculate the fraction of A in each mixture and enter into the table.

**26.** Which liquid do you think is more dense, liquid A or liquid B? Explain your answer.

**27.** Use Logger Pro to construct a correctly labeled graph of density vs. moles of B.

**28.** Do a linear fit of this data, and print the graph. Is the graph consistent with a pure liquid A density of 1.05 g/mL? Explain.

**29.** Use Logger Pro to construct a correctly labeled graph of density vs. fraction of A.

**30.** Do a linear fit of this data, and print the graph. Does this graph predict the density of pure liquid B? If so, what is the density of pure liquid B. Explain how you arrived at this answer.

**31.** What is the density of a mixture that is 2.3 moles A and 2.7 moles B? Explain how you arrived at this answer.

# IUPAC Inorganic Nomenclature

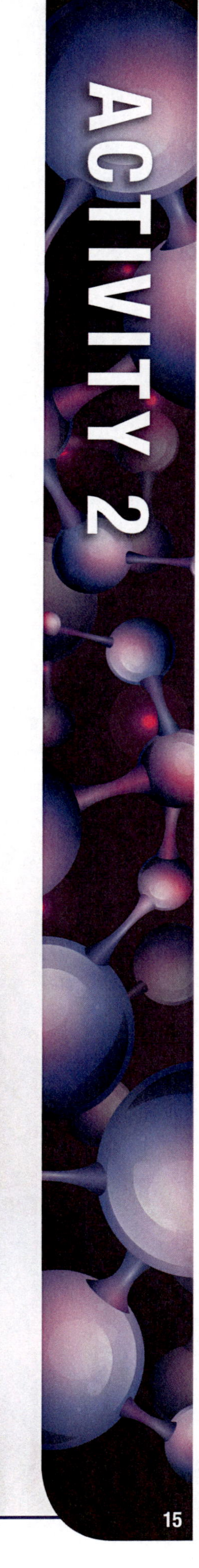

# 20-QUESTION PRE-LAB ASSIGNMENT

CRN:______________________________________     Your Name: ______________________________

Date Submitted: _________________________     TA's Name: ______________________________

*Please write your answers legibly in the space provided. Some answers will actually consist of three or four pieces of information that are clearly written on the board during the conversation **or** clearly stated verbally during the presentation. You must include all aspects of the answer to receive full credit. These answers are due at the beginning of the lab meeting.*

**1.** What is this week's topic?

**2.** What does the word **nomenclature** mean?

**3.** What is required for a substance to be designated **organic?**

**4.** What does IUPAC mean? And what do they do?

**5.** List the five classes of chemical compounds.

**6.** How are metals separated from non-metals on the periodic table?

**7.** What do the prefixes **mon**(o) and **poly** mean in the words **monatomic** and **polyatomic?**

**8.** Define **cation** and **anion.**

**9.** What are the group/family ionic charges on the periodic chart? (List the top most family member for each group, and write the family ionic charge above it.)

**10.** We used five specific examples—one for each class of chemical compound. First, write the chemical symbol or the polyatomic ion used in each example.

   **a.** M & NM:

   **b.** M & PA:

   **c.** PC & PA:

   **d.** PC & NM:

   **e.** NM & NM:

**11.** Next, place a charge behind each M, NM, PC, or PA, and then show the result of the "criss-cross."

   **a.** M   & NM   :

   **b.** M   & PA   :

   **c.** PC   & PA   :

   **d.** PC   & NM   :

   **e.** NM   & NM   :

**12.** Finally, draw out the ions as spheres for the first two classes.

    **a.** M & NM:                                          **b.** M & PA:

**13.** What is the meaning of the word **formula unit?**

**14.** Show how to get to the simplest whole number ratio for $CaCO_3$. What do we call this simplest whole number ratio?

**15.** What is the empirical formula for $Hg_2I_2$? Why? (Draw the formula unit as part of your answer.)

**16.** List the nomenclature rules.

**17.** Name our first four examples of compounds.

**18.** What is the subtlety for chemical compounds that contain only **non-metals?**

**19.** What are the prefixes we use for each number 1–10?

**a.** 1:

**b.** 2:

**c.** 3:

**d.** 4:

**e.** 5:

**f.** 6:

**g.** 7:

**h.** 8:

**i.** 9:

**j.** 10:

**20.** For metals with multiple possible charges, like $Cu^{+1/+2}$, what can we use to designate the charges of +1 and/or +2? Show how we constructed the chemical formulas **and** name the two copper sulfates.

# BACKGROUND

| TABLE 1 | |
|---|---|
| **Polyatomic Cations** | |
| Ammonium | $NH_4^+$ |
| Hydronium | $H_3O^+$ |
| Mercury(I), Hg(I) | $Hg_2^{2+}$ |

| TABLE 2 | |
|---|---|
| **Polyatomic Anions** | |
| Acetate | $C_2H_3O_2^-$ |
| Azide | $N_3^-$ |
| Bicarbonate | $HCO_3^-$ |
| Bisulfate | $HSO_4^-$ |
| Bisulfite | $HSO_3^-$ |
| Carbonate | $CO_3^{2-}$ |
| Chlorate | $ClO_3^-$ |
| Chlorite | $ClO_2^-$ |
| Chromate | $CrO_4^{-2}$ |
| Cyanide | $CN^-$ |
| Dichromate | $Cr_2O_7^{2-}$ |
| Dihydrogen Phosphate | $H_2PO_4^-$ |
| Hydroxide | $OH^-$ |
| Hypochlorite | $ClO^-$ |
| Monohydrogen Phosphate | $HPO_4^{2-}$ |
| Nitrate | $NO_3^-$ |
| Nitrite | $NO_2^-$ |
| Oxalate | $C_2O_4^{2-}$ |
| Perchlorate | $ClO_4^-$ |
| Permanganate | $MnO_4^-$ |
| Peroxide | $O_2^{2-}$ |
| Phosphate | $PO_4^{3-}$ |
| Phosphite | $PO_3^{3-}$ |
| Sulfate | $SO_4^{2-}$ |
| Sulfite | $SO_3^{2-}$ |
| Thiocyanate | $SCN^-$ |
| Thiosulfate | $S_2O_3^{2-}$ |

**TABLE 3**

| Metals with Multiple Possible Charges | |
|---|---|
| Copper | $Cu^+$ and $Cu^{2+}$ |
| Iron | $Fe^{2+}$ and $Fe^{3+}$ |
| Cobalt | $Co^{2+}$ and $Co^{3+}$ |
| Manganese | $Mn^{2+}$ and $Mn^{7+}$ |
| Mercury | $Hg(I)$ and $Hg^{2+}$ |
| Tin | $Sn^{2+}$ and $Sn^{4+}$ |
| Lead | $Pb^{2+}$ and $Pb^{4+}$ |
| Bismuth | $Bi^{3+}$ and $Bi^{5+}$ |
| Arsenic | $As^{3+}$ and $As^{5+}$ |

# DATA SHEET

Name: _______________________________     Grade: _______________________________

Date Experiment Performed: _______________     Days Late: _______________________

CRN of Lab Section: _______________________     Instructor's Initials: _______________

## Writing Empirical Formulas and Naming Ionic Compounds, Acids, and Hydrates

Using the "criss-cross" method, write valid empirical formulas for the compounds that would be formed by combining the given atoms and polyatomic ions. Then name the compound.

If a chemical formula is provided, name that chemical compound.

For example, sodium with chlorine:

$Na^+$ & $Cl^-$

NaCl

Sodium chloride

**1.** Combine

   **a.**

   **b.**

   **c.**

   **d.**

   **e.**

   **f.**

   **g.**

   **h.**

i.

j.

k.

l.

m.

n.

o.

p.

q.

r.

s.

t.

u.

v.

w.

x.

y.

z.

aa.

ab.

ac.

ad.

ae.

af.

ag.

ah.

ai.

aj.

ak.

al.

am.

**2.** Provide the correct chemical formula for the following IUPAC names

    **a.** Acetic acid:

    **b.** Boron trihydride:

    **c.** Hydroiodic acid:

    **d.** Carbon tetrachloride:

    **e.** Iodine heptafluoride:

    **f.** Phosphorus pentabromide:

    **g.** Tetraphosphorus decasulfide:

    **h.** Calcium sulfate dihydrate:

    **i.** Cobalt(II) chloride hexahydrate:

    **j.** Barium peroxide octahydrate:

    **k.** Methane:

# Numismatists' Delight
## How Many Drops of Water Will the Surface of a Penny Accommodate



## Objective

At the completion of the lab, you should be able to

- determine exactly how many drops of water will fit on a penny that is lying flat on a flat surface.

Will the same number of drops fit on both heads and tails? We shall see!

# PROCEDURE

## Materials

- a penny
- a ruler
- the smallest beaker in your lab drawer
- a disposable pipette

1. Obtain a 1¢ coin—Lincoln cents are recommended, but Canadian pennies, German pfennig, etc., may all be used. Please note the country of origin of your penny. ***Both partners should use the exact same coin.***

2. Record the date of your penny.

3. If you use a Lincoln cent, please also report what is on the reverse (the tails) side of the coin—the wheat stalks, the Lincoln Memorial, a scene from Abe's life, the shield, etc.

4. If you use a penny other than a Lincoln cent, please describe what is on the obverse (the heads) side as well as on the reverse side.

5. Measure and report the diameter of your penny in both inches (in) and centimeters (cm).

6. Place a small "square" of paper towel or a kimwipe on the bench top.

7. Lay your penny flat on the absorbent paper.

8. Fill your beaker approximately ⅔ full of distilled water from the DI tap.

9. Fill your disposable pipette with distilled water from your beaker.

10. Very slowly and very deliberately, add one drop of water at a time to the surface of your penny.

11. Continue to add drops of water to the surface of your penny, ***and keep count*** of how many drops you can add before the water flows over the "rim" of your penny. (That last drop that causes the water to gush over the rim should not be reported in your count.)

12. Note whether or not the water's surface seemed to "bulge" over the sides of the penny.

13. Draw a sketch of the water on your penny from both a map view perspective (looking down from the penny above) as well as from a profile perspective (sometimes described as "edge on").

14. Both lab partners should count, in three separate trials each, the number of drops of water that the obverse side can hold before the water spills over the rim; ***and***

15. Count, in three separate trials each, the number of drops of water that the reverse side can hold before the water spills over the rim.

16. Fill in Table 1, and bring these data with you next week so that we can use them for the Math Review Workshop.

# DATA SHEET

Name: _______________________________     Grade: _______________________________

Date Experiment Performed: _______________     Days Late: _______________________________

CRN of Lab Section: _______________________     Instructor's Initials: _______________________

| TABLE 1 | | |
| --- | --- | --- |
| Coin Used | U.S. "Lincoln" Penny | |
| Date on Coin | | |
| Obverse Shows | Abraham Lincoln's Profile | |
| Reverse Shows | | |
| Diameter (inches) | | |
| Diameter (centimeters) | | |
| | **Your Number of Drops** | **Your Partner's Number of Drops** |
| Obverse Trial 1 | | |
| Obverse Trial 2 | | |
| Obverse Trial 3 | | |
| Reverse Trial 1 | | |
| Reverse Trial 2 | | |
| Reverse Trial 3 | | |

A few questions to ponder …

1. What are some liquids other than water that you are aware of? That you have had personal experience with? List at least seven liquids other than water that you know about or have personally encountered.

**2.** Would the same number of drops of a different liquid "fit" on the penny?

**3.** Why do you think this?

**4.** In what ways do all liquids behave like water?

**5.** In what ways do different liquids behave differently from water?

# Measurements and Density



## Objectives

At the completion of the lab, you should be able to

- record data and take notes in a laboratory notebook so that the information can be used later to analyze the data;

- use a top-loading balance properly to determine the mass of a material;

- use and read a burette properly to measure the volume of a liquid;

- use and read a graduated cylinder to measure the volume of a liquid;

- select a measurement tool that is appropriate for a particular measurement;

- make a graph that plots the mass and volume of a liquid;

- prepare a solution of a specific proportion;

- perform dilutions of solutions and calculate the resulting concentrations;

- describe the concepts of density, relative density, and specific gravity;

- calculate density and specific gravity; and

- decide how to best determine the density of an object and perform the density measurements.

# 20-QUESTION PRE-LAB ASSIGNMENT

CRN:_______________________________________    Your Name: _______________________________

Date Submitted: ________________________________    TA's Name: _______________________________

*Please write your answers legibly in the space provided. Some answers will actually consist of three or four pieces of information that are clearly written on the board during the conversation **or** clearly stated verbally during the presentation. You must include all aspects of the answer to receive full credit. These answers are due at the beginning of the lab meeting.*

1. What is the **in**tensive physical property studied in this lab experiment? What is the definition of "**in**tensive"?

2. What is special about a **balance?**

3. What are the three steps for using a balance?

4. What should we know about the zero at the end of the measurement 9.990 g?

5. What are the three devices used to measure liquid volume in this experiment, their smallest graduation, and the place you will estimate to (and report)?

**6.** What gives rise to **uncertainty?**

**7.** What is the typical uncertainty in a measurement? Give the volume example and the mass example shown in the video.

**8.** What are the two types of solids we can obtain volume for?

**9.** How do we obtain the volume of a regular solid?

**10.** Who was the scientist that gave us the method for determining the density of irregular solids? What is that method called?

**11.** And what was his famous utterance? What does that word translate into?

**12.** How do we calculate the volume of an irregular solid using Archimedes' principle? (Draw the labeled graduated cylinder as part of your answer.)

**13.** Show the calculation we did in the video to determine the percent of students wearing EKU apparel in class that day.

**14.** What is the equation we will use to calculate percent by mass, % mass?

**15.** To how many places behind the decimal point should you report *all* your % mass answers in this lab (even if it violates significant figure rules)?

**16.** Write down our general equation for the % mass of salt calculation. And then show the sample calculation we did in the video for Salt Solution I, including the final answer reported.

**17.** Sketch the two beakers and show the ingredients as we go from Beaker I to Beaker II.

**18.** Show the density calculation we did in the video for Solution I.

**19.** Reproduce the sketches of the eggs in the beakers that we can use to help us to determine the relative density of an egg.

**20.** What does the term "relative" mean in "relative density"?

# BACKGROUND

One of the important skills needed by a chemist is the ability to make accurate and precise measurements and to know which measurement tool is most appropriate for different situations.

In this experiment you will use a variety of devices to measure mass, volume, and length. Then you will apply this knowledge to determine the density of several different items.

You should record all of your data and observations *in your laboratory notebook.* You will use that information later to write up the report for the lab.

## Materials

**If you are unfamiliar with any of this equipment, please refer to the Blackboard post under the Course Documents tab that has pictures of common chemistry lab equipment on it.**

- balance
- penny
- 250-mL beaker
- 100-mL graduated cylinder
- 10-mL graduated cylinder
- burette
- burette clamp and ring stand
- test tube
- ruler
- metal cylinders
- rubber stoppers
- 400-mL or 600-mL beakers (four of these for each pair of students)
- sodium chloride
- distilled water
- hard-boiled egg (one for each pair of students)
- plastic spoon (one for each pair of students)

# PROCEDURE

*Be sure to record all of your observations and data in your lab notebook. You will need many of these observations to complete the lab report.*

## Part I: Use of Balance to Measure the Mass of an Object

A balance is used in a laboratory to determine mass. (Remember that a scale does not determine mass—it determines weight.) You will be using an electronic top-loading balance. The blue cover over the balance pan protects against drafts and dust accumulation.

### Determine the Mass of the Penny at Your Lab Station

1. Check the balance pan and be sure that it is clean.
2. Turn the balance on.
   a. How many decimal places can you read with the balance in this lab?
   b. What is the capacity of the balance?
3. Place a piece of filter paper or weighing paper on the balance pan. Notice that the weighing container does have some mass.
4. To account for the mass of the weighing container, push the "zero" or "tare" button to zero the balance.
5. When the display reads 0.00(0) grams, place the penny on the weighing container and determine its mass. Record the mass of the penny. Don't forget to include the unit of measurement. **Record every single digit that the balance provides.**

## Part II: Volume Measurements

There are many tools in the lab that can be used to measure volume, including eye droppers, beakers, Erlenmeyer flasks, graduated cylinders, burettes, and pipettes. These devices differ in the accuracy and precision of the measurement. For example, an eye dropper would neither be very accurate nor precise.

The calibration marks on these volume-measuring devices are different. Make sure that you know the numerical meaning of the marks. You must *estimate* one more digit beyond the graduations on the device when you are recording data. For example, if the device is calibrated to show 0.1-mL volumes, you must *estimate* the reading to 0.01 mL.

Also remember that, because of water's surface tension, it tends to cling to the sides of glass cylinders, forming a curved water surface called a *meniscus.* You should read the volume at the **bottom** of the meniscus; your eye should be at meniscus level.

In Figure 1.1, each line stands for 0.1 mL. So, you could *estimate* one more decimal and record a result to the hundredths place.

The volume of the liquid in the picture to the right would be recorded as 7.78 mL. The last digit is estimated. Be sure that you understand how this reading is obtained.

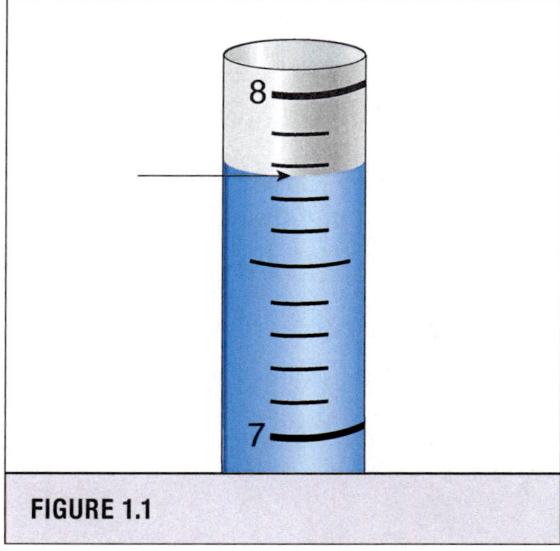

**FIGURE 1.1**

## Examination of Measuring Devices and Measurement of Volume

Obtain the following items and line them up on your lab bench:

- 250-mL beaker
- 100-mL graduated cylinder
- 10-mL graduated cylinder
- burette

***Record the following information in your lab notebook:**

1. Look at the 250-mL beaker at your lab desk.

   a. There are calibration lines on the 250-mL beaker. Make a simple sketch of these calibration lines.

   b. Does the beaker read volume from the bottom up or from the top down?

   c. To what place could you record (the recorded measurement includes the estimated place) using a 250-mL beaker for volume measurement?

2. Look at the 100-mL graduated cylinder at your lab desk.

   a. There are calibration lines on the 100-mL cylinder. Describe these lines.

   b. What volume is represented by each line?

   c. Does the cylinder read volume from the bottom up or from the top down?

   d. To what place could you record (the recorded measurement includes the estimated place) using a 100-mL cylinder for volume measurement?

3. Look at the 10-mL (or 25-mL) graduated cylinder at your lab desk.

   a. There are calibration lines on the 10-mL cylinder. Describe these lines.

   b. What volume is represented by each line?

   c. To what place could you record (the recorded measurement includes the estimated place) using a 10-mL cylinder for volume measurement?

4. Look at the burette.

   a. Describe the calibration lines on the burette.

   b. Does the burette read volume from the bottom up or from the top down?

   c. What volume is shown by the smallest calibration marks imprinted on the burette?

   d. To what place could you record (the recorded measurement includes the estimated place) using a burette for volume measurement?

## Measurement of Volume

There will be a test tube in your lab drawer. We want to determine the volume of water that can fit in this test tube.

Use all four pieces of glassware for volume measurement to do this. Be sure that you record each volume with the correct number of sig. figs. for each measuring device. (This recorded measurement includes the final place that you estimate.)

- beaker (use the 50-mL beaker in your lab drawer)
- 100-mL graduated cylinder
- 10-mL graduated cylinder
- burette

Fill the test tube up to the very top with distilled water, and then pour that into the beaker and record the measurement using the graduations on the beaker.

Fill the test tube up to the very top with distilled water, and then pour that into the 100-mL graduated cylinder and record the measurement using the graduations on the 100-mL graduated cylinder.

Fill the test tube up to the very top with distilled water, and then pour that into the 10-mL graduated cylinder and record the measurement using the graduations on the 10-mL graduated cylinder.

Fill the burette to the 0.00-mL mark with distilled water, and then place the test tube under the burette tip. Deliver water from the burette until the test tube is completely full. Read and record the measurement of the water dispensed from the burette using the graduations on the burette.

# Part III: Using Measuring Devices to Determine the Density of Solids

Density is a physical property that determines the mass of a substance in a specific volume.

$$D = \frac{mass}{volume}$$

The volume of a solid can be determined either by length measurements (if it has a regular shape) or by water displacement (if it has an irregular shape).

*For the following sections, remember to record all data and a brief description of what you did in your lab notebook.*

## Density of a Rubber Stopper

Determine the density of the #3, 2-hole rubber stopper available at your lab desk.

## Density of a Metal Cylinder

Determine the density of the piece of metal available at your lab desk.

# Part IV: Determination of the Density of Water by Graphing

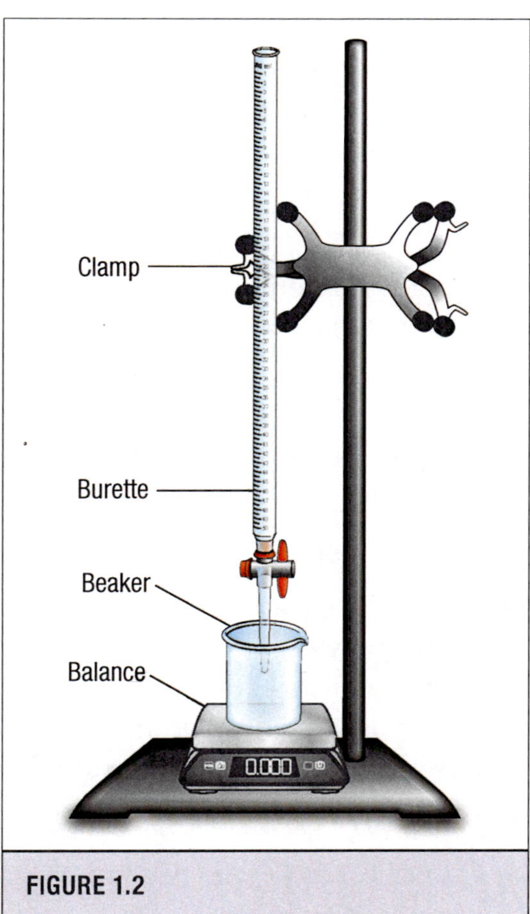


**FIGURE 1.2**

1. Get a clean burette and fill it nearly to the top of the markings with distilled water. Hold the burette over the sink or a beaker, and let water drain out of it until the water level falls below the top calibrated part of the burette. Set the burette up in the clamp **over the balance** as shown in the diagram. Leave room for a 50-mL beaker. Record the initial volume of water in the burette in your notebook.

2. Turn on the balance. Place a 50-mL beaker on the balance, and zero the balance.

3. Open the stopcock and deliver **exactly** 10.00 mL of water into the beaker. After this has been added, record the mass of the water.

4. Add another 10.00 mL of water to the beaker, and record the mass.

5. Refill your burette to the "zero" graduation.

6. Add another 10.00 mL of water to the beaker, and record the mass.

7. Add another 10.00 mL of water to the beaker, and record the mass.

You should now have four mass readings and four volume readings.

Using the graph paper provided in lab, construct a graph with the mass of the water on the $y$-axis and the volume on the $x$-axis. Remember to label all axes, title the graph, put your name and date on it, and scale it properly. Each partner should make their own copy of the graph.

# Part V: Determination of Relative Density of a Hard-Boiled Egg

The relative density of a substance is sometimes a useful quantity. This is simply the density of a material relative to the density of some other material. For most liquids and solids, the standard material is water. For your egg, it will be the density of the egg relative to the density of the salt solution. For the solution between the ones where the egg floats well and it sinks all the way to the bottom, the density of the egg is the density of that intermediate solution.

In this part of the experiment, you will use the balance and graduated cylinders to prepare a *stock* solution of sodium chloride. You will then *dilute* this stock solution to make a series of solutions of varying density. By observing whether an egg floats or sinks, you will get an idea about the relative density of a hard-boiled egg.

***Record all data and observations in your lab notebook.**

1. Obtain a piece of weighing paper, fold it half once, and then in half again perpendicular to the first fold. Unfold the piece of weighing paper, and place it on the balance in the center of the balance pan with the folds opening up. Tare the balance to 0.0 g with the weighing paper on it.

2. Add about 50 grams of sodium chloride (table salt) into the center of the weighing paper. **Record the exact mass you used.** (Record every single digit that the balance displays.)

3. Pour the sodium chloride into a 400-mL (or larger) beaker. Use the 100-mL graduated cylinder (filled to the 100.0-mL mark and then emptied three times) to add 300.0 mL of water to the 400-mL beaker. Measure the 100.0 mL as accurately as possible each of the three times.

4. Stir well, until the salt has dissolved.

5. There are hard-boiled eggs in the cooler on one of the side lab tables. Put the egg into this first salt solution. Note its position in the solution. (Sink, suspend, or float? Does it float high, does it hover between the surface of the solution and the bottom of the beaker, or does it sink completely and rest on the bottom of the beaker?) **Record your observation(s) with a diagram in your lab notebook.**

6. Remove the egg with a spoon and set it aside.

7. Take 150.0 mL of the first salt solution you made and put it into a second 400-mL (or larger) beaker. Add 150.0 mL of distilled water to the second beaker and mix well.

8.  Put the egg into Solution 2. Note its position in the solution. (Sink, suspend, or float? Does it float high, does it hover between the surface of the solution and the bottom of the beaker, or does it sink completely and rest on the bottom of the beaker?) **Record your observation(s) with a diagram in your lab notebook.**

9.  Remove the egg with a spoon and set it aside.

10. Take 150.0 mL of the second salt solution you made, and put it into a third 400-mL (or larger) beaker. Add 150.0 mL of distilled water and mix well.

11. Put the egg into Solution 3. Note its position in the solution. (Sink, suspend, or float? Does it float high, does it hover between the surface of the solution and the bottom of the beaker, or does it sink completely and rest on the bottom of the beaker?) **Record your observation(s) with a diagram in your lab notebook.**

12. Remove the egg with a spoon and set it aside.

13. Take 150 mL of the third salt solution you made, and put it into a fourth 400-mL (or larger) beaker. Add 150 mL of distilled water and mix well.

14. Put the egg into Solution 4. Note its position in the solution. (Sink, suspend, or float? Does it float high, does it hover between the surface of the solution and the bottom of the beaker, or does it sink completely and rest on the bottom of the beaker?) **Record your observation(s) with a diagram in your lab notebook.**

15. Remove the egg with a spoon and set it aside.

16. Now put the egg into a beaker that contains only distilled water. Note its position in the solution. (Sink, suspend, or float? Does it float high, does it hover between the surface of the solution and the bottom of the beaker, or does it sink completely and rest on the bottom of the beaker?) **Record your observation(s) with a diagram in your lab notebook.**

# Part VI: Determination of the Density of the Hard-Boiled Egg

Determine the density of the hard-boiled egg directly by measuring its mass and volume. In your notebook, describe how you did this.

**Turn in a copy of your lab notebook pages to your TA before you leave lab today!**

## Clean-Up and Disposal

- The salt solutions can be poured down the sink drains.
- Rinse the hard-boiled egg, dry it, and return it to the cooler.
- Rinse out all of the beakers you used, and put them back into the lab drawer.
- Return all of the items used for density determinations to the box on the lab bench.
- Be sure that any spills are brushed off the balance and that the balance is turned off.
- Wipe up all of the area around your lab bench.

## Report and Due Date

The blank lab report form is available on the Blackboard site. Reports are due *at the beginning of the next lab period.*

# DATA SHEET

Name: _________________________________     Grade: _______________________________

Date Experiment Performed: _______________     Days Late: ___________________________

CRN of Lab Section: ______________________     Instructor's Initials: _________________

| General Grading Items | 20 Points |
| --- | --- |
| Copies of Lab Pages Submitted; Labeled with Name and Date, Complete Information, Readable, Data Recorded Matches Results Given in Report | |
| Name and CRN Included on Report | |
| All Safety Rules Were Followed | |
| Waste Was Properly Disposed of and Lab Area Was Cleaned | |
| Evaluation of Student Performance Overall (Student Was on Time, Followed Safety Rules, Performed the Lab Correctly and Within the Time Allowed, Etc.) | |
| 20-Question Pre-Lab Assignment | /20 |
| **Total** | /20 |

**CHE 111L student: Submit the copies of your lab notebook pages for this experiment to your TA before you leave lab for the day.**

## Part I: Determination of Mass (3 points)

1. What was the mass of the penny?

2. What is the capacity of the balance you used?

3. How many decimal places can you read on the balance you used?

## Part II: Volume Determinations (12 points)

4. To what place can you record using a 250-mL beaker? (This recorded measurement includes the estimated place.)

5. To what place can you record using a 100-mL graduated cylinder? (This recorded measurement includes the estimated place.)

6. To what place can you record using a 10-mL graduated cylinder? (This recorded measurement includes the estimated place.)

7. To what place can you record using a burette? (This recorded measurement includes the estimated place.)

8. Volume of the test tube determined by pouring the full test tube into the beaker:

9. Volume of the test tube determined by pouring the full test tube into the 100-mL graduated cylinder:

10. Volume of the test tube determined by pouring the full test tube into the 10-mL graduated cylinder:

11. Volume of the test tube determined by filling it with water dispensed from the burette:

12. Suppose that you need to measure the volume of a liquid with the highest accuracy and precision possible. Which of these volume measuring devices used in the experiment would you choose?

13. Suppose that you needed a quick, approximate measurement of about 200 mL of water. Which measuring device would you choose?

14. Record the correct reading for the volume measurement in Figure 1.3: and in Figure 1.4:

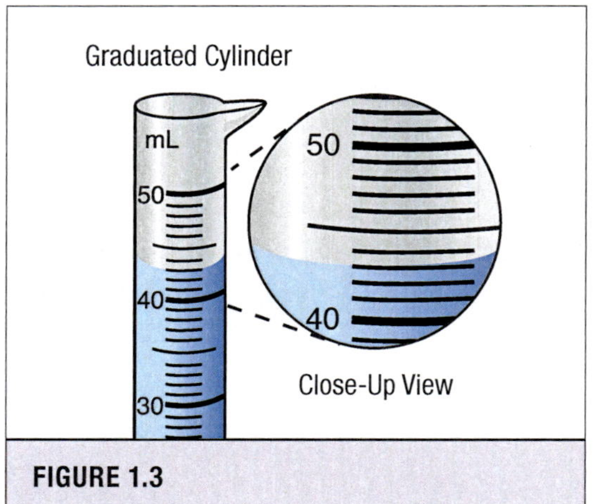


**FIGURE 1.3**

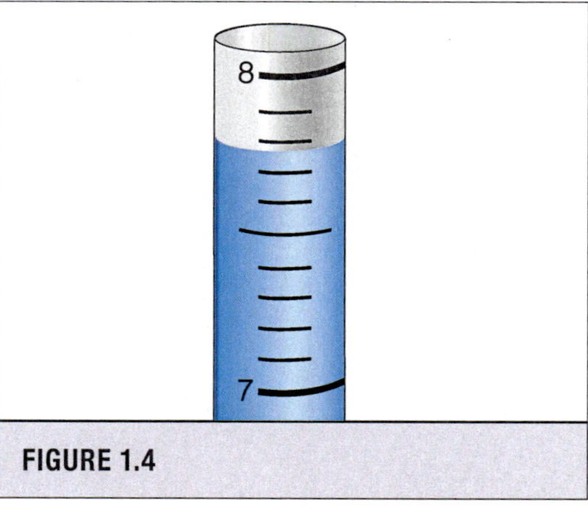


**FIGURE 1.4**

# Part III: Density of Solids (10 points)

**15.** What was the density of the rubber stopper? Show the mass and volume you used, and show the result with units and correct significant figures.

How did you determine the volume of the stopper?

**16.** What was the density of piece of metal? Show the mass and volume you used, and show the result with units and correct significant figures.

How did you determine the volume of the metal?

**17.** What are the two densest naturally occurring elements, and what is the density of each?

**18.** What is the lowest density solid, and what is the value of its density?

## Part IV: Density of Water (9 points)

**19.** Attach the graph you made plotting the mass and volume of the water.

**20.** What is the physical meaning of the slope of this graph? What is the numerical value of the slope?

**21.** What is the density of water, **according to your graph?**

## Part V: Relative Density of a Hard-Boiled Egg (46 points)

**22.** Calculate the percentage salt and the density of each of the four sodium chloride solutions you made. ***Show the numbers you used to calculate your answer. And report four sig. figs. for each answer!***

Also note whether the egg floated or sank in each solution.

**Record your observation(s) of the egg floating, being suspended, or sinking to the bottom of the beaker with a diagram in your lab notebook.**

   **a. Solution 1**

Does the egg sink, suspend, or float?

**b. Solution 2**

Does the egg sink, suspend, or float?

**c. Solution 3**

Does the egg sink, suspend, or float?

**d. Solution 4**

Does the egg sink, suspend, or float?

**23.** What was the relative density of the hard boiled egg? Explain how you got your answer.

**24.** Use references to find out the approximate salinity (in weight %) of the Dead Sea. Would the egg float in this water?

**25.** Would a human body be expected to float in the Dead Sea? Please explain.

**26.** How did you **directly measure** the density of the egg? What was your measured density? Show the mass and volume you used for your answer.

**27.** Finally, compare your directly measured density to the approximate **relative** density you got from the salt solutions. Are the two values about the same? Different? If different, can you suggest a way to account for those differences?

# The Determination of the Chemical Formula of a Hydrated Salt

## Objectives

At the completion of the lab, you should be able to

- make accurate and precise mass determinations with the balance;
- learn proper set-up of a crucible and clay triangle for heating;
- heat a substance to constant mass; and
- use mole concepts and stoichiometry to determine the correct chemical formula for two hydrated compounds.

# 20-QUESTION PRE-LAB ASSIGNMENT

CRN:_________________________________________     Your Name: _______________________________

Date Submitted: ____________________________     TA's Name: _______________________________

*Please write your answers legibly in the space provided. Some answers will actually consist of three or four pieces of information that are clearly written on the board during the conversation **or** clearly stated verbally during the presentation. You must include all aspects of the answer to receive full credit. These answers are due at the beginning of the lab meeting.*

  **1.** What exactly is a "salt"?

  **2.** What does "hydrated" mean?

  **3.** What is "laundry bluing"?

  **4.** What is the correct IUPAC name and the correct chemical formula for laundry bluing?

  **5.** Which ratio is this experiment attempting to discover? What are some typical values for this ratio?

  **6.** The prefix **"an-"** means _________________ , and so **an**aerobic" means _________________ , while **"an**hydrous" means _________________ .

  **7.** What is the ratio of interest for laundry bluing?

**8.** What is that ratio for the hydrated salt of nickel(II) sulfate?

**9.** Define "stoichiometry."

**10.** From the definition for stoichiometry, what are "the basics" in chemistry?

**11.** Why do we use a clay triangle to support the crucible during heating and cooling?

**12.** Draw a sketch of the crucible and the lid during heating. Why is the lid ajar?

**13.** Draw and label a sketch of **flame etiquette** below.

**14.** Why must the crucible be cool before weighing?

**15.** How many different salts will you dehydrate in lab?

**16.** Although the lines of data that you will record on the Data Sheet may seem to be out of sequential order, there is actually a reason that we record the data in a different order than they get reported on the form. What is the first measurement that you make, and which line does it get reported on? What is the second measurement that you make, and which line is it reported on?

**17.** Many measurements that a scientist makes in lab are done "approximately accurately." What does it mean to make a measurement "approximately accurately"? (You may use the example of measuring mass from the video or the task of measuring approximately accurately 10 mL of water from a burette like you did in Experiment 1.)

**18.** How many times must you heat your salt sample?

**19.** What must the difference between two consecutive weighings ("massings") be less-than-or-equal-to in order for us to accept that all the water that can be driven off has been driven away from the sample?

**20.** Describe the process of "normalization" and what we expect the numerical result to be.

# BACKGROUND

*Hydrates* are compounds that have a specific number of water molecules attached to them. A hydrate is shown by writing the formula for the compound and then an elevated dot followed by the number of water molecules attached. For example, copper(II) sulfate crystals have five water molecules for every copper(II) sulfate molecule. The prefix for five is *penta,* so we would name this copper(II) sulfate pentahydrate. The formula would be $CuSO_4 \cdot 5\ H_2O$. (Please note that the dot between the salt formula and the number designation for the waters of hydration is right smack in the middle of the line. It is not a "period.")

The molar mass of copper(II) sulfate pentahydrate would include the five water molecules, and would be 249.5 g/mol. If you need to, be sure to review how this value is obtained.

The water of hydration can often be driven off of a solid by heating it. In this experiment, you will be given two different hydrated ionic compounds and asked to determine the formula for the compounds with the water of hydration. You will know the identity of the salt, but not how many waters of hydration are present. You will need to take very careful mass measurements to get acceptably good ("accurate") results for this experiment. Use balances that report to the 1000ths place.

## Materials

- approximately 1 gram of a hydrate sample, measured accurately
- clean crucible and cover
- crucible tongs
- burner
- ring stand, ring, and clay triangle

# PROCEDURE

1. Obtain a crucible and cover.

2. Rinse the crucibles well with water. (Tap water is okay.)

3. Thoroughly wipe and *dry* the crucible. Then **determine the mass of the crucible (and cover) to ±0.001 g.** Record all of the data on the Data Sheet.

   Once you have cleaned the crucible, don't handle it any more with your fingers because fingerprint grease can add to the crucible mass. Use crucible tongs to handle the crucible.

4. Get approximately 1 g of a hydrate sample in a small test tube.

   **After you obtain your sample, quickly replace the cap to the reagent bottle and tighten securely.**

5. Add the hydrate to the crucible. It will be okay if not all of the hydrate is transferred.

6. Record the mass of the sample, crucible, and cover to ±0.001 g. Remember that the hydrate doesn't have to be exactly 1.000 g. Accurately record the actual mass.

7. Set up a ring stand with a ring and clay triangle, similar to that shown in Figure 2.1. Using your crucible tongs, place the crucible, cover, and contents on the clay triangle. Partially cover the opening of the crucible with the cover. Leave the cover in a position that will let the water of hydration escape as you heat the sample.

   Adjust the height of the ring so that it is about 2 inches above the top of the burner.

8. Light the burner and adjust so that you have a **blue** flame (*not a yellow one; soot will be deposited on the crucible and increase the mass*). Move the burner under the crucible and begin to heat the crucible and its contents for about 5 minutes. ***Do not allow the crucible to turn red.*** **(Overheating may lead to decomposition of your sample!)**

**FIGURE 2.1**

9. Turn off the burner. Use the tongs and put the cover all the way over the crucible. Remove the crucible to a cooling pad, and allow the crucible to cool to room temperature. (You can begin work on the second sample while you are waiting for the first one to cool.) Be patient—it will take some time for the crucible to cool to room temperature. Remember not to touch the crucible to see if it is cool. You can just hold your hand very close to the crucible to test.

10. After the crucible is cool, determine the mass of the crucible and contents (and cover) to ±0.001 g. *Use the same balance you used to obtain the initial masses.* Record on the Data Sheet.

11. Calculate the mass of the salt after the first heating and record on the Data Sheet.

12. We want to be sure that we have driven off all of the water of hydration, so we will repeat the heating process until we get two "massings" in-a-row that agree with each other to within 0.03 g.

13. Put the crucible back on the clay triangle and heat again using a medium blue flame for about 5 minutes. Leave the cover slightly ajar as before.

14. Turn off the burner, put the top all the way on, and allow the crucible to cool to room temperature as before. (You may allow the crucible to cool in place on the triangle.)

15. After the crucible is cool, determine the mass of the crucible and contents (and cover) to ±0.001 g. *Use the same balance you used to obtain the initial masses.* Record on the Data Sheet.

16. Calculate the mass of the salt after the second heating and record on the Data Sheet.

17. Compare the mass of the salt after the first and second heating. The difference between the two masses should be **no more than 0.03 grams.** If the difference is more than this, heat the sample up again one more time. Let it cool and weigh it again. If the difference is still more than 0.03 g, repeat the heating and cooling process.

    This process of heating a material until its mass is constant is called *heating to constant mass.*

18. Repeat this process for a second, different hydrated compound.

## Clean-Up and Disposal

After you have heated your crucible and contents to constant mass and recorded the final mass, you can wash the crucible out. These hydrates can all go down the sink.

## Sample Data Set

| TABLE 2.1 | |
|---|---|
| Name/Formula on the Jar of the Salt You Used | $NiSO_4 \cdot n\, H_2O$ |
| Mass Crucible, Top, and Hydrated Salt **before** Heating | 26.162 g |
| Mass Crucible and Top Only | 25.080 g |
| Mass of Hydrated Salt before Heating | |
| Mass of Crucible, Top, and Salt after First Heating | 25.679 g |
| Mass of Salt Only after First Heating | |
| Mass of Crucible, Top, and Salt after Second Heating | 25.676 g |
| Mass of Salt Only after Second Heating | |
| Mass of Crucible, Top, and Salt after Third Heating, If Needed | 25.676 g |
| Mass of Salt Only after Third Heating, If Needed | |
| Mass of Crucible, Top, and Salt after Fourth Heating, If Needed | |
| Mass of Salt Only after Fourth Heating, If Needed | |
| Mass of Water Removed from the Salt | |

# DATA SHEET

Name: ___________________________________   Grade: ___________________________________

Date Experiment Performed: _________________________   Days Late: _______________________________

CRN of Lab Section: _______________________________   Instructor's Initials: _______________________________

| General Grading Items | 20 Points |
|---|---|
| Copies of Lab Pages Submitted; Labeled with Name and Date, Complete Information, Readable, Data Recorded Matches Results Given in Report | |
| Name and CRN Included on Report | |
| All Safety Rules Were Followed | |
| Waste Was Properly Disposed of and Lab Area Was Cleaned | |
| Evaluation of Student Performance Overall (Student Was on Time, Followed Safety Rules, Performed the Lab Correctly and Within the Time Allowed, Etc.) | |
| 20-Question Pre-Lab Assignment | /20 |
| **Total** | **/20** |

# Salt 1

## Data (15 points)

| TABLE 2.2 | |
| --- | --- |
| Name/Formula on the Jar of the Salt You Used | |
| Mass Crucible, Top, and Hydrated Salt *before* Heating | |
| Mass Crucible and Top Only | |
| Mass of Hydrated Salt before Heating | |
| Mass of Crucible, Top, and Salt after First Heating | |
| Mass of Salt Only after First Heating | |
| Mass of Crucible, Top, and Salt after Second Heating | |
| Mass of Salt Only after Second Heating | |
| Mass of Crucible, Top, and Salt after Third Heating, If Needed | |
| Mass of Salt Only after Third Heating, If Needed | |
| Mass of Crucible, Top, and Salt after Fourth Heating, If Needed | |
| Mass of Salt Only after Fourth Heating, If Needed | |
| Mass of Water Removed from the Salt | |

## Calculations (25 points)

Show work for all calculations.

1. Convert the mass of water removed to moles of water present in the hydrate. (6 points)

2. Convert the mass of the dehydrated salt to moles of salt. (Remember that this is the *dehydrated* salt—so use the correct molecular mass.) (6 points)

**3.** From the calculated results of Questions 1 and 2 above, determine the whole number mole ratio of water to salt in this molecule. (6 points)

**4.** Write the chemical formula for the hydrate and its IUPAC name, **based on *your* results.** (7 points)

# Salt 2

## Data (15 points)

| TABLE 2.3 | |
| --- | --- |
| Name/Formula on the Jar of the Salt You Used | |
| Mass Crucible, Top, and Hydrated Salt *before* Heating | |
| Mass Crucible and Top Only | |
| Mass of Hydrated Salt before Heating | |
| Mass of Crucible, Top, and Salt after First Heating | |
| Mass of Salt Only after First Heating | |
| Mass of Crucible, Top, and Salt after Second Heating | |
| Mass of Salt Only after Second Heating | |
| Mass of Crucible, Top, and Salt after Third Heating, If Needed | |
| Mass of Salt Only after Third Heating, If Needed | |
| Mass of Crucible, Top, and Salt after Fourth Heating, If Needed | |
| Mass of Salt Only after Fourth Heating, If Needed | |
| Mass of Water Removed from the Salt | |

## Calculations (25 points)

Show work for all calculations.

**1.** Convert the mass of water removed to moles of water present in the hydrate. (6 points)

**2.** Convert the mass of the dehydrated salt to moles of salt. (Remember that this is the *dehydrated* salt—so use the correct molecular mass.) (6 points)

**3.** From the calculated results of Questions 1 and 2 above, determine the whole number mole ratio of water to salt in this molecule. (6 points)

**4.** Write the chemical formula for the hydrate and its IUPAC name, **based on *your* results.** (7 points)

# Solution Conductivity
## Electrolytes, Dissociation, and Ionization



## Objectives

At the completion of the lab, you should be able to

- define conductivity, electrolyte, and ion;
- distinguish among strong, weak, and non-electrolytes;
- distinguish between ionization and dissociation;
- determine the conductivity of a solution using a conductivity probe;
- use measured conductivity values to predict whether a substance is a strong, weak, or non-electrolyte; and
- filter solutions using vacuum filtration.

## Reference

Silberberg and Amateis, 8e; Section 4.1, *Solution Concentration and the Role of Water as a Solvent,* pages 146–149, and *Dilution of Solutions,* pages 152–154.

# 20-QUESTION PRE-LAB ASSIGNMENT

CRN:_________________________________     Your Name: _________________________________

Date Submitted: _____________________     TA's Name: _________________________________

*Please write your answers legibly in the space provided. Some answers will actually consist of three or four pieces of information that are clearly written on the board during the conversation **or** clearly stated verbally during the presentation. You must include all aspects of the answer to receive full credit. These answers are due at the beginning of the lab meeting.*

**1.** In lab, what kind of water do we use? What are the other two types of water mentioned?

**2.** Define "solution" and "homogeneous."

**3.** Distinguish **solvent vs. solute**—in general and for Kool-Aid and brine.

**4.** What is the special name for a solution with water as the solvent?

**5.** What types of compounds can dissolve in water?

**6.** What combines to form salts? to make molecular compounds?

**7.** Name and describe the two steps of the **dissolution process for a salt,** and then write the process equation for the dissolution of table salt.

**8.** Name and describe the one step that occurs for the **dissolution of molecular compounds,** and then write the process equation for the dissolution of table sugar.

**9.** Draw the **spheres of hydration** for $Na^+(aq)$, $Cl^-(aq)$, and sucrose$(aq)$.

**10.** Acids are **molecular compounds.** Do they dissolve in a manner that is more like salts *or* more like other molecular compounds? Write the process equation for the dissolution of $HNO_3$.

**11.** When discussing the electrical conductivity of solutions, the "old" (and still reliable!) classification scheme of **ionic vs. covalent** gives way to a new classification scheme. Name the two "new" categories and the name types of compounds in each.

**12.** Define "electrolyte."

**13.** Define "non-electrolyte."

**14.** Contrast **strong electrolytes vs. weak electrolytes.**

**15.** Provide two examples of strong and two examples of weak electrolytes.

**16.** What is the standard SI unit for electrical conductivity? What is the unit for electrical conductivity that we use in this lab?

**17.** List the chemical formula for acetic acid in four different ways.

**18.** Why do we list H first in CHE 111L when writing the chemical formula for an acid?

**19.** Write the dissociation/hydration process equation for the dissolution of a weak electrolyte like acetic acid. Pay special attention to write the proper "arrow."

**20.** Define "precipitate." How can we recognize that a precipitate has formed in lab?

# BACKGROUND

In this experiment you will measure the conductivity of a solution to help you better understand the concepts of ionic and molecular compounds, what electrolytes are, and the difference between dissociation and ionization.

You know that the main difference between molecules and ions is that ions have a charge, while molecules are neutral. For example, chloride ion ($Cl^-$) has a negative charge, and sodium ions ($Na^+$) have a positive charge. A molecule of carbon dioxide ($CO_2$) does not have any charge, nor does a molecule of sugar ($C_{12}H_{22}O_{11}$).

The movement of charged particles constitutes an electric current. If there are no charged particles that are free to move, there won't be any current flow. The *conductivity* of a substance or solution is a measure of how well it conducts electricity. For example, you know that metals will conduct electricity, while substances such as rubber and plastics will not conduct electricity. An easy way to determine the conductivity of a substance is to use a light bulb attached to two metal strips that are separated by a few centimeters. If the two metal strips are put into a conducting solution, the light bulb will light up. If the solution does not conduct electricity, the light bulb will not light up.

Although this is an easy way to see whether or not a solution conducts electricity, this kind of *conductivity apparatus* has limited usefulness because it can't tell us how well the solution conducts electricity. A better way to determine conductivity is with a meter that will provide us with actual conductivity values. In this experiment, we will use a special type of probe to perform this measurement. Conductivity is measured in units of microsiemens/cm.

An *electrolyte* is a substance that forms an aqueous solution that conducts electricity better than pure water. Any substance that forms ions when placed into water will be an electrolyte. And, the more ions that are present, the higher the conductivity will be. A *nonelectrolyte* will not conduct electricity.

## Materials

- beakers of various sizes
- side-arm flask with hose
- filter paper to fit suction funnel
- suction funnel
- straws

## Reagents

- sodium carbonate (solid), $Na_2CO_3$
- calcium chloride (solid), $CaCl_2$
- 0.1 M sodium chloride solution, $NaCl(aq)$
- 1 M acetic acid solution, $HC_2H_3O_2(aq)$
- 1 M ammonium hydroxide solution, $NH_4OH(aq)$
- 0.1 M hydrochloric acid solution, $HCl(aq)$
- 0.1 M sodium hydroxide solution, $NaOH(aq)$

# PROCEDURE

## Part I: Set-Up the Conductivity Probe

In this experiment you will use a Vernier unit with conductivity probe for the measurements.

1. Locate a Vernier LabPro unit. The unit, probe, and cables should be in a plastic box on the top of the lab bench (Figure 3.1).

2. Find the conductivity probe (Figure 3.2). Look at the side of the probe, and you will see that it has three possible settings. For most of this experiment, you will use the 0–20000 µS/cm setting. Set the switch to that setting (Figure 3.3).

3. The conductivity probe has a cord with a white connector (Figure 3.4). This connector should be plugged into the CH 1 connector on the side of the Vernier unit. The other end of the cord on the conductivity unit is attached to the probe.

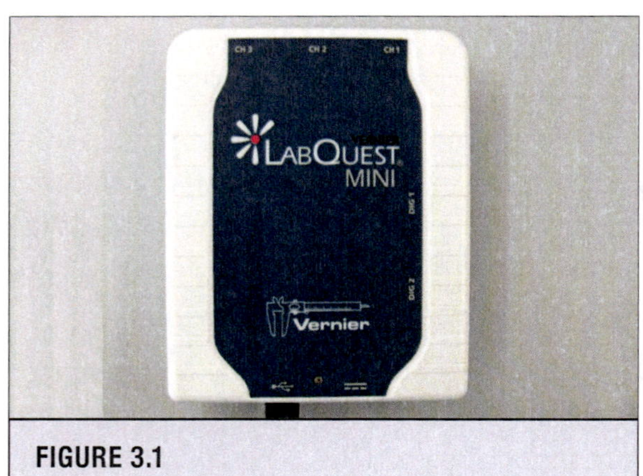

FIGURE 3.1

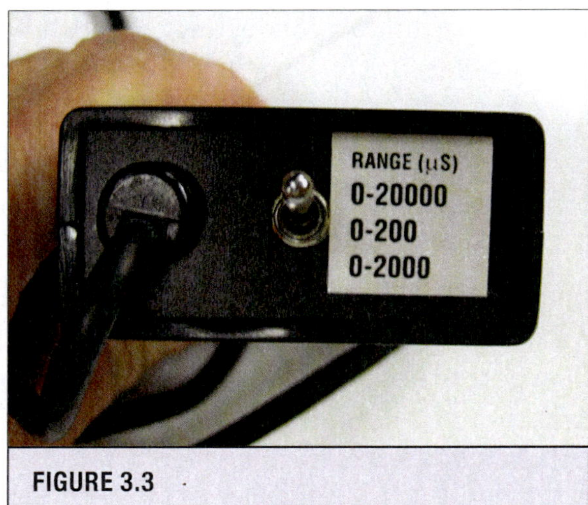


FIGURE 3.3

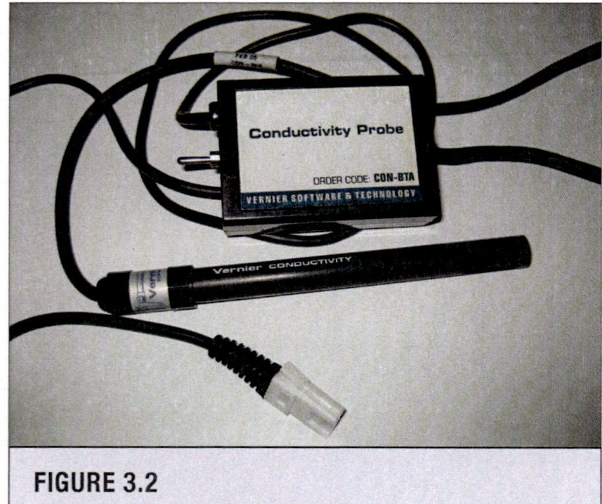


FIGURE 3.2

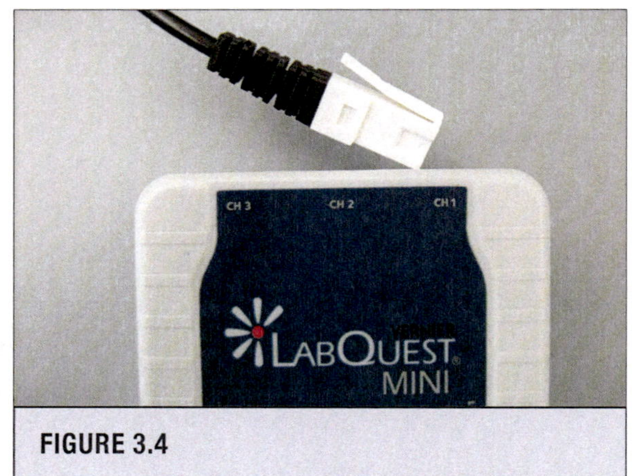

FIGURE 3.4

4. Locate the cord with a USB connection on one end and a printer-type connector on the other end (Figure 3.5). The printer-type connector goes into the Vernier unit on the side. The USB connection goes to the computer.

5. Plug in the Vernier unit to electrical power. You should hear some beeps and see a green light on the unit when it is ready.

6. After everything is connected, log into the computer and open Logger Pro. Logger Pro should recognize the probe automatically. The conductivity reading will appear in red in the lower left corner of the computer monitor.

7. To zero the probe, click "experiment" from the top menu bar, then select "zero."

8. When you use the probe, be sure that the metal piece in the tip of the probe is submerged in the solution (Figure 3.6).

9. For all of the experiment except the testing of distilled water and double distilled water, you should have the switch set to 0–20000 µS/cm. For the distilled waters, set to 0–200 µS/cm. You will get a message on the screen asking you to re-zero when you change settings.

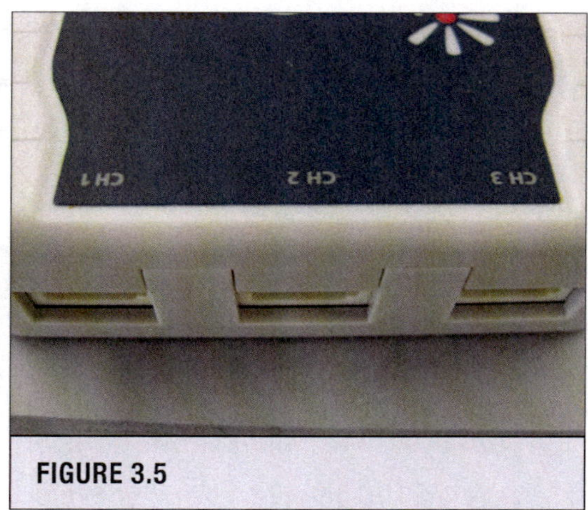

**FIGURE 3.5**

**FIGURE 3.6**

# Part II: Conductivity of Water and Ionic Compounds

10. Test the conductivity of each of the materials below. Use a small beaker and put enough liquid in the beaker to cover the metal portion of the probe (Figure 3.7). Record the conductivity reading in your notebook.

   **Be sure to rinse the probe well with distilled water between each solution in all parts of the experiment.**

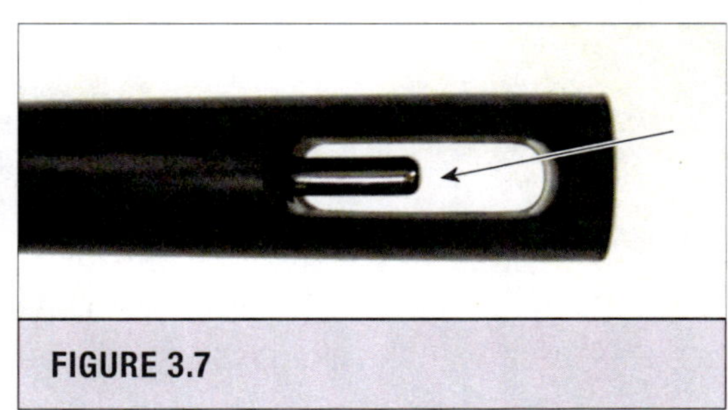

**FIGURE 3.7**

   a. double distilled water (set switch to 0–200 µS/cm)

   b. distilled water (set switch to 0–200 µS/cm)

   c. tap water (set switch to 0–20000 µS/cm for all remaining tests)

   d. 0.01 M NaCl (Make this by diluting the 0.1 M NaCl by a factor of 10)

**Note:** *Be sure to use the **dilution equation** $C_1V_1 = C_2V_2$. For example, to make 50 mL of 0.01 M NaCl from the 0.1M NaCl:*

$$(0.1 \text{ M NaCl})(V_1) = (0.01 \text{ M NaCl})(50 \text{ mL}).$$

Solve for $V_1$, and you will get 5 mL. So, you would take 5 mL of the 0.1 M NaCl and add 45 mL of water.

11. Use weighing paper and weigh 0.5–0.6 g individual samples of the two salts listed below. Record the masses in your notebook.

    a. $Na_2CO_3$ (sodium carbonate)

    b. $CaCl_2$ (calcium chloride)

12. Get two 100-mL clean, dry beakers. Add one sample to each beaker, and then add 50 mL of distilled water to each beaker. Stir until completely dissolved. Describe the appearance of each solution in your lab notebook.

13. Test the conductivity of each salt solution and record.

14. Get a 250-mL beaker and combine the total volumes of these salt solutions and mix thoroughly.

15. Describe any obvious change which occurred on preparation of this mixture. In your notebook, write the equation for the reaction which occurred. (Refer to solubility rules if necessary.)

16. Use vacuum filtration to filter the mixture you just made to separate the liquid portion from the precipitate. (Your TA will demonstrate this for you in lab.)

    a. Get a ring stand, a clamp, a filter flask with a side-arm on it, a vacuum hose to attach to the side-arm, a funnel that fits into the flask, and a piece of filter paper that fits into the funnel.

    b. Put the paper inside the funnel, and place the funnel into the flask.

    c. Clamp the flask in place on a ring stand so that it doesn't flip over when the heavy vacuum hose is attached.

    d. Attach the vacuum hose to the side-arm of the flask and to the vacuum aspirator nozzle (the spigot with "VAC" on the yellow dot) at each lab station.

    e. Turn on the vacuum valve. This will create a vacuum in the flask, and the solution will filter much faster. Slowly and carefully pour the solution to be filtered into the funnel. Don't pour in so much that it overflows.

    **f.** After you have filtered the entire sample, turn off the vacuum aspirator. Remove the funnel from the flask, and pour the filtrate (the liquid portion that collected in the flask) into a beaker and save it for the next part of the experiment. Replace the funnel on the flask.

    **g.** Wash the precipitate on the filter paper with a few mL of distilled water. This wash solution can be discarded.

**17.** Transfer as much of the white solid as you can from the filter paper to a clean beaker. Add 50 mL of distilled water and stir vigorously for at least one minute.

    **a.** Test the conductivity of this mixture and record.

    **b.** Identify this substance, based on your knowledge of solubility and components of the mixture you made.

    **c.** Comment on the water solubility of the substance.

**18.** Test and record the conductivity of the filtrate.

# Part III: Conductivity Testing of Acids and Bases

**19.** In separate clean beakers, obtain about 25 mL of each of the following four solutions. Be sure to mark the beakers so you will know which is which.

    **a.** 1 M acetic acid, $CH_3COOH$; also represented as HOAc

    **b.** 1 M ammonium hydroxide, $NH_4OH$

    **c.** 0.1 M hydrochloric acid, HCl

    **d.** 0.1 M sodium hydroxide, NaOH

**20.** In your lab notebook, make a species inventory for each one of these solutions.

**21.** Test each solution with a piece of litmus paper. Record your observations.

**22.** Test the conductivity of each of these solutions and record the results in your lab notebook. Keep the solutions for use in the next part.

**23.** Thoroughly mix the 1 M HOAc with the 1 M $NH_4OH$. Test the conductivity of the resulting solution and record the result in your notebook.

**24.** Test the mixture with a piece of litmus paper and record your observations.

**25.** Thoroughly mix the 0.1 M HCl with the 0.1 M NaOH. Test the conductivity of the resulting solution and record the result in your notebook.

**26.** Test the mixture with a piece of litmus paper and record your observations.

## Part IV: Conductivity Testing of "the Essence" of Human Breath in Water

Name the gas that is exhaled by humans as we breathe:

This same gas is also present in carbonated beverages.

27. Retest the conductivity of about 50 mL of doubled-distilled water to confirm that pure water is a very poor conductor of electricity. Remember to set the switch to 0–200 µS/cm for water readings. Record the conductivity value.

28. Perform a litmus paper test on the double distilled water. Record the results.

29. Pour 20 mL of "fizzy" 7-Up into the water and stir thoroughly. Test the conductivity of this solution. Record in your notebook.

30. Perform a litmus paper test on the water after you have added the 7-Up to it. Record the results.

31. Prepare 50 mL of 0.1 M $NH_4OH$ by diluting the 1 M ammonium hydroxide solution. (See note above about dilutions and an example of using the "dilution equation.") Mix the diluted solution thoroughly and divide it in half—i.e., pour ~25 mL of the 0.1 M $NH_4OH$ into a graduated cylinder or small beaker and set it aside for use in Step 36 below.

32. Perform a litmus paper test on the 0.1 M ammonium hydroxide solution. Record the result.

33. Test the conductivity of the 0.1 M ammonium hydroxide solution. Record the result in your notebook.

34. Pour 20 mL of the "fizzy" 7-Up into the 0.1 M $NH_4OH$ sample and stir thoroughly. Retest the conductivity of this solution and record.

35. Perform a litmus paper test on the 0.1 M ammonium hydroxide after you have poured the "fizzy" 7-Up into it. Record the results.

36. Repeat Steps 29–35 for the "flat" 7-Up provided.

## Clean-Up and Disposal

- All aqueous waste may be disposed in the sink.

- All solid wastes may be disposed in the trash.

- Return the Vernier unit, along with the probe and all cables, to the plastic box on the bench top.

# DATA SHEET

Name: _______________________________     Grade: _______________________________

Date Experiment Performed: _______________________     Days Late: _______________________________

CRN of Lab Section: _______________________     Instructor's Initials: _______________________

| General Grading Items | 20 Points |
|---|---|
| 20-Question Pre-Lab Assignment | /20 |
| | |
| | |
| | |
| **Total** | /20 |

**CHE 111L student: Remember to turn in a copy of your raw data notebook pages to your TA before you leave for the day.**

# Part II: Conductivity of Water and Ionic Compounds (30 points)

**1.** Summarize your conductivity measurements for the following materials

    **a.** Distilled water:                         **c.** Double distilled water:

    **b.** Tap water:                                **d.** 0.01 M NaCl:

**2. Based on your conductivity measurements,** compare the concentration of ions in the solutions (which has most ions and which has fewest?)

    **a.** Solution with most ions:

    **b.** Solution with 2nd highest number of ions:

    **c.** Solution with 3rd highest number of ions:

    **d.** Solution with fewest ions:

**3.** Masses of salts used

    **a.** $Na_2CO_3$:                               **b.** $CaCl_2$:

**4.** Equation for dissociation of sodium carbonate:

**5.** Equation for dissociation of calcium chloride:

**6.** Measured conductivity of each solution

    **a.** $Na_2CO_3$:                                   **b.** $CaCl_2$:

**7.** Description of changes that occurred after mixing the sodium carbonate and calcium chloride solutions:

**8.** Write the molecular equation for the reaction that took place upon mixing. Be sure to note any precipitates formed (remember solubility rules from the Experiment 4).

**9.** Conductivity of the solution you made from the white solid on the filter paper.

**10.** What is the white solid?

**11.** How soluble is calcium carbonate? Use your data from the conductivity measurement to support your answer.

**12.** Conductivity of the filtrate:

**13.** What ions were present in the filtrate?

# Part III: Conductivity of Acids and Bases (30 points)

**14.** Record the conductivity results, the litmus paper color, and a species inventory for these solutions.

**TABLE 3.1**

| Substance | Conductivity | Litmus Color | Species |
|---|---|---|---|
| 1 M $HC_2H_3O_2$ | | | |
| 1 M $NH_4OH$ | | | |
| 0.1 M HCl | | | |
| 0.1 M NaOH | | | |

**15.** Conductivity and litmus color of the **mixture** of 1 M HAc with 1 M $NH_4OH$

    **a.** Conductivity:             **b.** Litmus color:

**16.** Write the molecular equation for the reaction that took place when you mixed the 1 M HAc with 1 M $NH_4OH$.

**17.** Is the conductivity of the mixture higher or lower than the conductivity of the original solutions? Explain why.

**18.** Conductivity and litmus color of the **mixture** of 0.1 M HCl with 0.1 M NaOH

   **a.** Conductivity:                 **b.** Litmus color:

**19.** Write the molecular equation for the reaction that took place when you mixed the 0.1 M HCl with 0.1 M NaOH.

**20.** Is the conductivity of the mixture more or less than the conductivity of the original solutions? Explain why.

# Part IV: Conductivity Testing of "the Essence" of Human Breath in Water

## Acid–Base Chemistry of "Fizzy" 7-Up (10 points)

**21.** Record the conductivity and litmus color of double distilled water.

   **a.** Conductivity:                 **b.** Litmus color:

**22.** Record the conductivity and litmus color of the water after adding "fizzy" 7-Up to it.

   **a.** Conductivity:                 **b.** Litmus color:

**23.** Write a reaction for the addition of carbon dioxide to water.

**24.** Was the conductivity of the water before and after adding the "fizzy" 7-Up to it different? Explain your answer.

**25.** Record the conductivity and litmus color of 0.1 M $NH_4OH$ before you added "fizzy" 7-Up to it.

   **a.** Conductivity:　　　　　　　　　　**b.** Litmus color:

**26.** Record the conductivity and litmus color of 0.1 M $NH_4OH$ after adding "fizzy" 7-Up to it.

   **a.** Conductivity:　　　　　　　　　　**b.** Litmus color:

**27.** Was the conductivity of the $NH_4OH$ before and after adding "fizzy" 7-Up into it different? Explain your answer.

## Acid–Base Chemistry of "Flat" 7-Up (10 points)

**28.** Record the conductivity and litmus color of double distilled water.

   **a.** Conductivity:　　　　　　　　　　**b.** Litmus color:

**29.** Record the conductivity and litmus color of the water after adding "flat" 7-Up to it.

   **a.** Conductivity:　　　　　　　　　　**b.** Litmus color:

**30.** Write a reaction for the addition of carbon dioxide to water.

**31.** Was the conductivity of the water before and after adding the "flat" 7-Up to it different? Explain your answer.

**32.** Record the conductivity and litmus color of 0.1 M $NH_4OH$ before you added "flat" 7-Up to it.

   **a.** Conductivity:                              **b.** Litmus color:

**33.** Record the conductivity and litmus color of 0.1 M $NH_4OH$ after adding "flat" 7-Up to it.

   **a.** Conductivity:                              **b.** Litmus color:

**34.** Was the conductivity of the $NH_4OH$ before and after adding "flat" 7-Up into it different? Explain your answer.

# Chemical Reactions in Aqueous Solutions

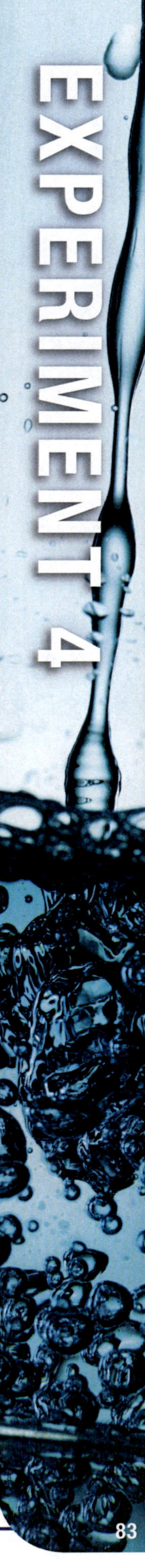


## Objectives

At the completion of the lab, you should be able to

- predict the products of a various displacement reactions;
- predict the products of a neutralization reaction;
- use a solubility table to predict the solublility of a particular compound in water;
- define precipitate and recognize which species would form a precipitate in a reaction;
- do a "species inventory" for a solution and list each species that would be present in a reaction mixture, along with an indication of whether it would be dissolved (*aq*) or a solid (*s*);
- perform chemical reactions and make complete and accurate observations about the reaction;
- write a balanced molecular equation for a chemical reaction;
- write a complete ionic equation for a chemical reaction; and
- write a net ionic equation for a chemical reaction.

## Reference

Silberberg and Amateis, 8e; selected portions of Sections 4.2 and 4.3, pages 155–160, as well as selected portions of Section 4.4, pages 165–171.

# 20-QUESTION PRE-LAB ASSIGNMENT

CRN:_______________________________________     Your Name: _______________________________

Date Submitted: _______________________________     TA's Name: _______________________________

*Please write your answers legibly in the space provided. Some answers will actually consist of three or four pieces of information that are clearly written on the board during the conversation **or** clearly stated verbally during the presentation. You must include all aspects of the answer to receive full credit. These answers are due at the beginning of the lab meeting.*

**1.** We say that a chemical reaction is a chemical change. What changes?

**2.** List the five clues that a chemical reaction has occurred.

**3.** Define "effervescence."

**4.** Why might odor change and/or temperature change be difficult for someone to detect in CHE 111L lab?

**5.** What is the name of the first type of chemical reaction presented in the pre-lab video?

**6.** Write a general reaction equation for it.

**7.** Today we are specifically studying                         replacement.

**8.** What is the name of the second type of chemical presented in the pre-lab video?

**9.** This reaction type is further subdivided into two categories. List them below.

**10.** Write the general equation for this second type of reaction below.

**11.** What are the three types of chemical equations that can be written?

**12.** In the space below, write all three types of chemical equations for the combination of sodium carbonate solution and calcium chloride solution.

**13.** What is the special name given to ions like $Na^+$ and $Cl^-$ that are part of your answer to Question 12?

**14.** For a precipitation reaction, what does the net ionic equation describe?

**15.** Write the three types of chemical equations for the acid–base neutralization of cesium hydroxide with hydroiodic acid.

**16.** What does this net ionic equation describe?

**17.** Write out all three types of chemical equations for our (balanced) single replacement example of aluminum plus sulfuric acid.

**18.** What are the two gases that **might** form when barium is dropped into water?

**19.** Name and describe the two "gas tests" we can perform to decide which of the two possible gases actually formed.

**20.** In the space provided, write out all three types of chemical equations for the reaction that occurs when barium is dropped into water.

# BACKGROUND

In lecture you learned that chemical equations are used to represent chemical reactions. In this lab you will be running several examples of "displacement" or "exchange" reactions in aqueous (water) solutions. In this type of reaction the ions appear to be "changing places." An example of an exchange, or double displacement reaction, is:

$$AgNO_3(aq) + NaCl(aq) \rightarrow NaNO_3(aq) + AgCl(s)$$

In this reaction it looks like the cations (the positively charged ions) silver and the sodium have switched places. That is why these are called *double displacement.* Remember that the symbol (*aq*) means that a substance is in water solution, and the symbol (*s*) means that a substance is in the solid phase.

An example of a single displacement reaction is:

$$Cu(s) + 2\ AgNO_3(aq) \rightarrow 2\ Ag(s) + Cu(NO_3)_2(aq)$$

## Solubility Rules

In order to decide whether or not a solid will be formed in a solution reaction, you need to know about the concept of solubility.

For this lab we will be using the following general solubility rules:

**TABLE 4.1**

| Ion(s) | Solubility in Water | Exceptions |
|---|---|---|
| Group IA Ions ($Li^+$, $Na^+$, $K^+$, $Rb^+$, $Cs^+$) | Soluble | None |
| Ammonium ($NH_4^+$) | Soluble | None |
| Group VIIA Ions—the Halides ($F^-$, $Cl^-$, $Br^-$, $I^-$) | Soluble | $Ag^+$, $Hg^{2+}$, $Pb^{2+}$ |
| Nitrates ($NO_3^-$) | Soluble | None |
| Acetates ($C_2H_3O_2^-$) | Soluble | None |
| Sulfates ($SO_4^{2-}$) | Soluble | $Ca^{2+}$, $Ba^{2+}$, $Sr^{2+}$, $Hg^{2+}$, $Pb^{2+}$, $Ag^+$ |
| Carbonates ($CO_3^{2-}$) | Insoluble | Group IA and Ammonium |
| Phosphates ($PO_4^{3-}$) | Insoluble | Group IA and Ammonium |
| Hydroxides ($OH^-$) | Insoluble | Group IA, $Ca^{2+}$, $Ba^{2+}$, $Sr^{2+}$ |
| Sulfides ($S^{2-}$) | Insoluble | Group IA and IIA Ions and Ammonium |

For example, is the compound CoS soluble in water? In Table 4.1, the last line tells you that sulfides are generally insoluble in water, unless the compound has Group I or II elements or ammonium ion in it. Cobalt is not in group I or II, and it isn't ammonium. So, CoS would **not** be soluble in water. Would $LiNO_3$ be soluble in water? The answer is yes. In the table above, you can see in the first row that all Group IA ions are water soluble, and in the fourth row you can see that all nitrates are soluble.

When you mix two solutions, you can often tell whether or not a reaction has occurred by looking for the formation of a *precipitate* (a solid that forms upon mixing, and abbreviated ppt) or looking for the formation of a gas. Sometimes you will see bubbles if a gas is formed. Sometimes you can smell the gas or see the gas if it has a color. For a flammable gas like hydrogen, you can do a "pop" test. For oxygen, you can do a "glowing splint" test. In this experiment, you will be looking for the formation of a precipitate and for indicators of gas formation.

## Neutralization Reactions

An important type of exchange reaction involves acids and bases. An acid and a base react to form salt and water in a *neutralization* reaction. There is usually no obvious sign that this neutralization has taken place. But, you could monitor the pH with litmus paper. You will learn more about the concept of pH in CHE 112L. For now, we will simply say that the pH is a number that tells us how acidic or basic a solution is. Litmus paper will turn red in acid and blue in base. A neutral solution won't result in a color change. An example of a neutralization reaction is:

$$HCl(aq) \quad + \quad NaOH(aq) \quad \rightarrow \quad NaCl(aq) \quad + \quad H_2O$$
$$\text{acid (red)} \qquad \text{base (blue)} \qquad \text{salt} \qquad \text{water (neutral)}$$

## Materials

- test tubes and rack
- glass stirring rod
- litmus paper
- matches
- zinc metal, small pieces
- magnesium metal, small pieces
- calcium metal, small pieces

## Solutions

- 0.1 M hydrochloric acid
- 0.1 M sodium hydroxide
- 0.1 M sodium chloride
- 0.1 M sodium bromide
- 0.1 M silver nitrate
- 0.1 M copper(II) sulfate
- 0.1 M iron(III) nitrate
- 0.1 M sodium iodide
- 6 M HCl

# PROCEDURE

You will run 10 reactions. For each reaction, you should take two test tubes to the dispensing station. Use the dispenser to get 1 mL of each of the two solutions listed for each reaction or a very small piece of the solid material if the reagent is a solid. Return to your lab station with the reagents. *In your lab notebook, record all of the information asked for below for each reaction.*

<table>
<tr>
<td>SAFETY!<br><br>WASTE DISPOSAL</td>
<td>When you have finished running each reaction, pour the contents of the tube into the large waste containers in the lab hood. Do not pour the contents down the sink. You can rinse the tubes well using regular tap water before going on to the next reaction. Rinse any leftover metal pieces with DI $H_2O$ and leave them to dry on a paper towel at your stations.</td>
</tr>
</table>

## Example of a Species Inventory

You will be asked to do a *species inventory* for the solutions. This is simply a list of all the different species present in the solution. Here is an example:

Suppose you are going to react magnesium chloride solution with silver nitrate solution. What species are actually present in each solution? Since both of these compounds are soluble in water, they would exist in water as ions. So, the species inventory would be shown in the diagrams below. In the magnesium chloride solution, magnesium ions, chloride ions, and water would be present. In silver nitrate, silver ions, nitrate ions, and water would be present. After reaction, magnesium and nitrate would still be present as ions because both are water soluble. However, silver chloride would have formed a solid precipitate. You know that by looking at the solubility tables and noting that although the halides are generally soluble, silver halides are an exception.

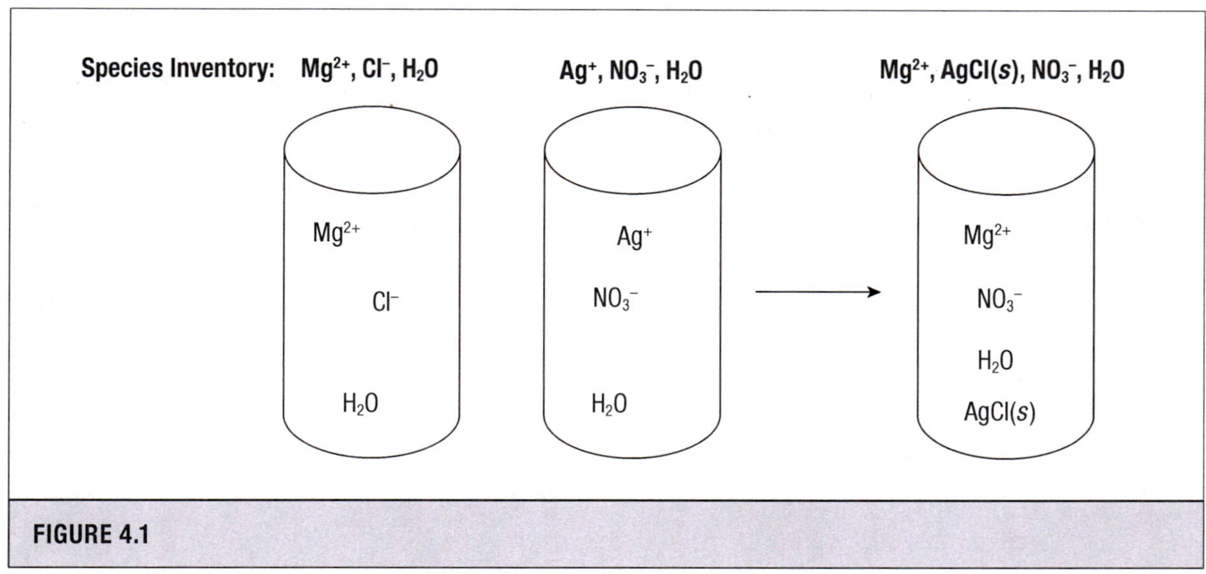


**FIGURE 4.1**

## Single Displacement Reactions (cation replacement)

1. Get a small piece of magnesium and 1 mL of 6 M hydrochloric acid in separate test tubes and take them back to your lab bench. Also, have a match ready to strike. Before running the reaction, make the following observations and record them *in your lab notebook.*

   a. Color of HCl solution

   b. Species inventory of HCl

   c. Appearance of magnesium metal

   d. Put the tube holding the Mg into a test tube rack. Add the HCl. Observe what happens. **Quickly** strike a match and put the match near the mouth of the test tube. Record your observations in your lab notebook.

2. Get a small piece of zinc and 1 mL of 6 M hydrochloric acid in separate test tubes, and take them back to your lab bench. Before running the reaction, make the following observations and record them *in your lab notebook.*

   a. Color of HCl solution

   b. Species inventory of HCl

   c. Appearance of zinc metal

   d. Put tube holding the Zn into a test tube rack. Add the HCl. Observe what happens. Quickly strike a match, and put the match near the mouth of the test tube. Record your observations in your lab notebook.

   e. Which reaction appeared more vigorous: the one with Zn or the one with Mg?

3. Get a small piece of zinc and 1 mL of *water* in separate test tubes, and take them back to your lab bench. Also, have a match ready to strike. Before running the reaction, make the following observations and record them *in your lab notebook.*

   a. Color of water

   b. Appearance of zinc metal

   c. Put the tube containing the zinc into a test tube rack. Add the water. Observe what happens. Strike a match and put the match near the mouth of the test tube. Record your observations in your lab notebook.

4. Get a small piece of calcium and 1 mL of *water* in separate test tubes, and take them back to your lab bench. Also, have a match ready to strike. Before reaction, make the following observations and record them *in your lab notebook.*

   a. Color of water

   b. Appearance of calcium metal

   c. Put the tube containing the calcium into a test tube rack. Add the water. Strike a match and put the match near the mouth of the test tube. Observe what happens. Record your observations in your lab notebook.

   d. Which reaction appeared more vigorous: the one with Zn or the one with Ca?

## Double Displacement Reactions (anion replacement)

5. Get 1 mL of silver nitrate and 1 mL of sodium chloride in separate test tubes, and take them back to your lab bench. Before running the reaction, make the following observations and record them *in your lab notebook.*

    a. Color of silver nitrate solution

    b. Species inventory for silver nitrate solution

    c. Color of sodium chloride solution

    d. Species inventory for sodium chloride solution

    e. Mix the silver nitrate and sodium chloride solutions. Shake the tube gently and mix the solutions thoroughly. Record your observations in your lab notebook.

6. Get 1 mL of silver nitrate and 1 mL of sodium bromide in separate test tubes, and take them back to your lab bench. Before running the reaction, make the following observations and record them *in your lab notebook.*

    a. Color of silver nitrate solution

    b. Species inventory for silver nitrate solution

    c. Color of sodium bromide solution

    d. Species inventory for sodium bromide solution

    e. Mix the silver nitrate and sodium bromide solutions. Shake the tube gently to mix the solutions thoroughly. Record your observations in your lab notebook.

7. Get 1 mL of silver nitrate and 1 mL of sodium iodide in separate test tubes, and take them back to your lab bench. Before running the reaction, make the following observations and record them *in your lab notebook.*

    a. Color of silver nitrate solution

    b. Species inventory for silver nitrate solution

    c. Color of sodium iodide solution

    d. Species inventory for sodium iodide solution

    e. Mix the silver nitrate and sodium iodide solutions. Shake the tube gently to mix the solutions thoroughly. Record your observations in your lab notebook.

## Double Displacement Reactions (cation replacement)

8. Get 1 mL of sodium hydroxide and 1 mL of copper(II) sulfate in separate test tubes, and take them back to your lab bench. Before running the reaction, make the following observations and record them *in your lab notebook.*

    a. Color of sodium hydroxide solution

    b. Species inventory for sodium hydroxide solution

    c. Color of copper(II) sulfate solution

    **d.** Species inventory for copper(II) sulfate solution

    **e.** Mix the copper(II) sulfate and sodium hydroxide solutions. Shake the tube gently to mix the solutions thoroughly. Record your observations in your lab notebook.

**9.** Get 1 mL of sodium hydroxide and 1 mL of iron(III) nitrate in separate test tubes, and take them back to your lab bench. Before running the reaction, make the following observations and record them *in your lab notebook.*

    **a.** Color of sodium hydroxide solution

    **b.** Species inventory for sodium hydroxide solution

    **c.** Color of iron(III) nitrate solution

    **d.** Species inventory for iron(III) nitrate solution

    **e.** Mix the iron(III) nitrate and sodium hydroxide solutions. Shake the tube gently to mix the solutions. Record your observations in your lab notebook.

## Neutralization

**10.** Get 1 mL of 0.1 M sodium hydroxide and 1 mL of 0.1 M hydrochloric acid in separate test tubes, and take them back to your lab bench. (**Note:** be sure to get 0.1 M HCl, not 6 M HCl.) Before running the reaction, make the following observations and record them *in your lab notebook.*

    **a.** Color of sodium hydroxide solution

    **b.** Species inventory for sodium hydroxide solution

    **c.** Color of hydrochloric solution

    **d.** Species inventory for hydrochloric acid solution

    **e.** Get three pieces of litmus paper, and put them on top of a piece of glass. Use the glass stirring rod and get a drop of the NaOH and put it on the litmus paper. Record the color of the litmus.

    **f.** Rinse the stirring rod well with water and get a drop of the HCl and put in on another piece of litmus paper. Record the color of the litmus.

    **g.** Rinse the stirring rod well with water and put a drop of distilled water and put in on the other piece of litmus paper. Record the color of the litmus.

    **h.** Mix the hydrochloric acid and sodium hydroxide solutions. Shake the tube gently to mix the solutions. Record your observations in your lab notebook.

    **i.** Test this solution on a piece of litmus paper and record your observations.

# Clean-Up and Disposal

- Clean the test tubes and other materials you used.
- Return everything to the drawer.
- Wipe the lab bench.

# DATA SHEET

Name: ___________________________________     Grade: _______________________________________

Date Experiment Performed: _____________________     Days Late: ________________________________

CRN of Lab Section: ___________________________     Instructor's Initials: ______________________________

| General Grading Items | 20 Points |
|---|---|
| Copies of Lab Pages Submitted; Labeled with Name and Date, Complete Information, Readable, Data Recorded Matches Results Given in Report | |
| Name and CRN Included on Report | |
| All Safety Rules Were Followed | |
| Waste Was Properly Disposed of and Lab Area Was Cleaned | |
| 20-Question Pre-Lab Assignment | /20 |
| **Total** | /20 |

**CHE 111L student: Submit copies of your lab notebook pages for this experiment to your TA before you leave for the day.**

# Single Displacement Reactions (cation replacement)

**1.** Mg and HCl (8 points)

    **a.** Color of HCl solution:          and species inventory:

    **b.** Description of Mg:

    **c.** Observations of reaction, including results of the pop test:

    **d.** If you got a positive pop test, what gas was present?

    **e.** Write the molecular equation for the reaction.

    **f.** Write the complete ionic equation for the reaction.

    **g.** Write the net ionic equation for the reaction.

**2.** Zn and HCl (8 points)

    **a.** Color of HCl solution:          and species inventory:

    **b.** Description of Zn:

**c.** Observations of reaction, including results of the pop test:

**d.** Which reaction appeared more active: HCl with Mg or HCl with Zn?

**e.** If you got a positive pop test, what gas was present?

**f.** Write the molecular equation for the reaction.

**g.** Write the complete ionic equation for the reaction.

**h.** Write the net ionic equation for the reaction.

**3.** Zn and Water (8 points)

**a.** Color of water:

**b.** Description of Zn:

**c.** Observations of reaction, including results of the pop test:

**d.** Which reaction appeared more active: Zn with HCl or Zn with water?

**e.** If you got a positive pop test, what gas was present?

**f.** Write the molecular equation for the reaction.

**g.** Write the complete ionic equation for the reaction.

**h.** Write the net ionic equation for the reaction.

**4.** Ca and Water (8 points)

**a.** Color of water:

**b.** Description of Ca:

**c.** Observations of reaction, including results of the pop test and whether the temperature of the test tube changed.

**d.** Which reaction appeared more active: water with Ca or water with Zn?

**e.** If you got a positive pop test, what gas was present?

**f.** Write the molecular equation for the reaction.

**g.** Write the complete ionic equation for the reaction.

**h.** Write the net ionic equation for the reaction.

# Double Displacement Reactions (anion replacement)

**5.** Silver nitrate and sodium chloride (8 points)

    **a.** Color of silver nitrate solution:      and species inventory:

    **b.** Color of sodium chloride solution:      and species inventory:

    **c.** Observations of reaction:

    **d.** Write the molecular equation for the reaction.

    **e.** Write the complete ionic equation for the reaction.

    **f.** Write the net ionic equation for the reaction.

**6.** Silver nitrate and sodium bromide (8 points)

    **a.** Color of silver nitrate solution:      and species inventory:

    **b.** Color of sodium bromide solution:      and species inventory:

    **c.** Observations of reaction:

    **d.** Write the molecular equation for the reaction.

    **e.** Write the complete ionic equation for the reaction.

    **f.** Write the net ionic equation for the reaction.

**7.** Silver nitrate and sodium iodide (8 points)

    **a.** Color of silver nitrate solution: _______ and species inventory: _______

    **b.** Color of sodium iodide solution: _______ and species inventory: _______

    **c.** Observations of reaction:

    **d.** Write the molecular equation for the reaction.

    **e.** Write the complete ionic equation for the reaction.

    **f.** Write the net ionic equation for the reaction.

# Double Displacement Reactions (cation replacement)

8. Copper(II) sulfate and sodium hydroxide (8 points)

    **a.** Color of copper(II) sulfate solution:      and species inventory:

    **b.** Color of sodium hydroxide solution:      and species inventory:

    **c.** Observations of reaction:

    **d.** Write the molecular equation for the reaction.

    **e.** Write the complete ionic equation for the reaction.

    **f.** Write the net ionic equation for the reaction.

9. Iron(III) nitrate and sodium hydroxide (8 points)

    **a.** Color of iron(III) nitrate solution:      and species inventory:

    **b.** Color of sodium hydroxide solution:      and species inventory:

    **c.** Observations of reaction:

**d.** Write the molecular equation for the reaction.

**e.** Write the complete ionic equation for the reaction.

**f.** Write the net ionic equation for the reaction.

# Neutralization

**10.** Hydrochloric acid and sodium hydroxide (8 points)

  **a.** Color of hydrochloric acid solution:                and species inventory:

  **b.** Color of litmus paper with a drop of HCl:

  **c.** Color of sodium hydroxide solution:                and species inventory:

  **d.** Color of litmus paper with a drop of NaOH:

  **e.** Color of litmus paper with a drop of distilled water:

  **f.** Observations of reaction:

  **g.** Color of litmus paper after the reaction:

  **h.** Did your litmus test show that the solution after reaction was acid, base, or neutral?

  **i.** Write the molecular equation for the reaction.

  **j.** Write the complete ionic equation for the reaction.

  **k.** Write the net ionic equation for the reaction.

# Solution Stoichiometry

## Objective

At the completion of the lab, you should be able to

- write a balanced chemical equation for the reaction between sodium hydroxide and acetic acid;
- use a volumetric pipette to accurately measure liquid volumes;
- perform stoichiometry calculations for a reaction to determine the concentration of an unknown;
- calculate the concentration of a solution in moles per liter;
- convert concentrations of moles per liter to % (weight/volume); and
- calculate the % error of your measurement.

# 20-QUESTION PRE-LAB ASSIGNMENT

CRN:_________________________________    Your Name: _______________________________

Date Submitted: _____________________    TA's Name: _______________________________

*Please write your answers legibly in the space provided. Many answers will actually consist of three or four pieces of information that are clearly written on the board during the conversation. You must include all aspects of the answer to receive full credit. Happy YouTubing!*

1. **Volumetric analysis** is also called ____________________________________.

2. Acid–base neutralization is a ____________________ type of reaction, and the general chemical equation for acid–base neutralization is written below:

3. Define **"titration."**

4. What does the phrase **"stoichiometric amount"** mean?

5. What *is* the **"equivalence point"**?

6. Write out and identify each term in the **titration equation.** What does the titration equation allow us to talk about quantitatively?

7. Write the general formula/equation for **concentration** and for the chemist's preferred concentration unit, **molarity.**

8. Sketch and label this week's **experimental set-up** as drawn on the chalkboard in the video.

**9.** What is the purpose of using the **magnetic stir bar?**

**10.** Write the chemical equation for the **acid–base titration** you will be performing this week.

**11.** What are the **three clues that a reaction has occurred?** Will the reaction represented by the chemical equation written above show any of these three clues?

**12.** What is the color of liquid water? What are the colors of the NaOH, acetic acid, and sodium acetate solutions?

**13.** What is a **chemical indicator?** How much of which chemical indicator do we use in our experiment mixture? What **happens to our chemical indicator** that allows us to "know" that the titration we perform has gone from start to finish?

**14.** How long must the fuchsia color persist?

**15.** In the three trials you run, how close to each other (in other words, how precise) must the volumes of base delivered be?

**16.** What is the **stoichiometry** of this balanced chemical equation?

**17.** Define (mass/volume) %, (m/v)%.

**18.** For the sample calculation, we were told that someone measured 10.38 mL of NaOH delivered from the burette in order to completely titrate the 10.00 mL of $HC_2H_3O_2$ solution in the Erlenmeyer flask. Please fill in the following information from that sample calculation.

10.38 mL of 1.000M NaOH =       moles of NaOH =       moles of $HC_2H_3O_2$

      moles of $HC_2H_3O_2$ =       g of $HC_2H_3O_2$ =       (m/v) % of $HC_2H_3O_2$,

which is       % (m/v) $HC_2H_3O_2$ to three significant figures.

**19.** Write the equations for calculating an **average** and a **standard deviation.**

**20.** We can obtain       from the **% error** calculation. Write that equation below, and report the % error for our sample calculation.

# BACKGROUND

During lecture you learned about the *stoichiometry* of reactions and about how to express the concentrations of solutions. This lab will require that you use your knowledge of both of these topics to run a reaction and perform some calculations to determine the concentration of an unknown solution.

You will react a solution containing an unknown amount of acetic acid with a solution containing a **known** concentration of sodium hydroxide. You will be told what this concentration is. You'll then use the known NaOH concentration and the stoichiometry of the reaction to figure out the concentration of the unknown acetic acid solution.

As you remember from lecture and the lab experiment on chemical reactions, the reaction of an acid with a base produces salt and water. What you'll do in this experiment is add *exactly enough* sodium hydroxide to react with the acetic acid. How will you know when you've added enough? The solution will turn a very pale pink because you will have added an *indicator* called phenolphthalein to the solution. You will learn more about how this process works in the second semester of general chemistry. For now, we are mainly interested in the reaction stoichiometry.

The concentration of a solution can be expressed in several ways. One of the most common units is *molarity,* which tells us the number of moles of solute per liter of solution. Another common way to express concentration is percent. The % (w/v) for a solution gives the number of grams of solute per 100 mL of solution. This is calculated with this formula:

$$\% \frac{w}{v} = \frac{\text{grams solute}}{100 \text{ mL solution}} \times 100\%$$

You will use both of these units in this experiment.

You will also use a *volumetric pipette* in this experiment. This is another device that can be used to measure volume. It is a very accurate and precise device for volume measurement.

## Materials

- acetic acid solution of unknown molarity
- sodium hydroxide solution with a known concentration (about 1 M)
- phenolphthalein indicator solution
- stirring plate with stand
- burette holder
- burette
- Ernlenmeyer flask
- beakers
- pH paper

## Clean-Up and Disposal

- The reaction of NaOH and acetic acid makes a salt (sodium acetate) and water. Both of these are safe to pour down the drain. You just need to be sure that there isn't excess acid or base present in the mixture.

- Get a large (400-mL or 600-mL) beaker from the glassware drawer. Use this beaker to collect all of the rinses and reaction mixtures you generate during the entire experiment.

- At the end of the experiment, get a piece of pH paper from the glassware drawer, and put it on a glass plate or hold it. Use a glass rod to get a drop of the solution from the beaker, and put it on the pH paper. As long as the pH of the solution is between 5 and 9 (weakly acidic to weakly basic), you can pour it down the drain. If the pH of the solution is below 5 or above 9, ask your instructor for assistance.

- Rinse the big beaker out well when you are finished, and return it to the glassware drawer.

# PROCEDURE

## Part I: Preparation of Burette with the Sodium Hydroxide

1. Get a burette, ring stand, burette clamp, and stir station if these are not already present on the bench top.

2. Close the stopcock on the burette. Fill the burette with distilled water, and drain it out into a beaker or into the sink. Discard this rinse water.

3. Get about 40 mL of the sodium hydroxide solution by dispensing it into a beaker. Note the concentration of the NaOH written on the container. Be sure to ***record the concentration of the sodium hydroxide solution in your lab notebook.*** Bring the NaOH back to your lab station.

4. Put the burette into the clamp, and put a small beaker under the burette. Close the stopcock. Use a plastic disposable pipette to draw up some NaOH. Put the tip of the pipette into the top of the burette, and let the NaOH run down all the sides of the burette to rinse it. Repeat this 2–3 times. Open the stopcock and let all of the rinse run into a beaker. The NaOH rinse should be put into the large beaker that you will use to collect all of your reaction material.

5. Close the burette stopcock. Add NaOH to the burette until it is above the zero mark. Put a beaker under the burette, and then drain some of the NaOH out of the burette until the meniscus falls within the calibration marks.

6. Check the tip of the burette to make sure there are no air bubbles in the tip. If there are, run out some more NaOH until the air bubbles have been removed.

7. **Record the initial reading for the NaOH in your lab notebook. You can make a table like the one below if you want to. Add columns for more trials if you need them.**

   **Remember to record the correct number of decimal places—you learned about this in Experiment 1 on measurements and density.**

| TABLE 5.1 | | | |
|---|---|---|---|
| | Trial 1 | Trial 2 | Trial 3 |
| Concentration of NaOH (molarity) | | | |
| Initial Volume of NaOH | | | |
| Final Volume of NaOH | | | |
| Volume of Unknown Acetic Acid | | | |

# Part II: Measurement of the Unknown Acetic Acid Using a Volumetric Pipette

8. Dispense about 50 mL of unknown acetic acid solution into a beaker and bring it back to your lab bench.

9. There will be a 10-mL volumetric pipette at your bench, along with a "thumbwheel" pipette pump. Notice that the volumetric pipette has only one mark on it for volume measurement. This kind of pipette has very good accuracy and precision, but it can only be used for a fixed volume.

10. Have a clean 125-mL Erlenmeyer flask (or larger) ready.

11. Insert the flat end of the volumetric pipette into the open end of the pipette pump. Be sure the wheel is turned all the way down on the pipette pump.

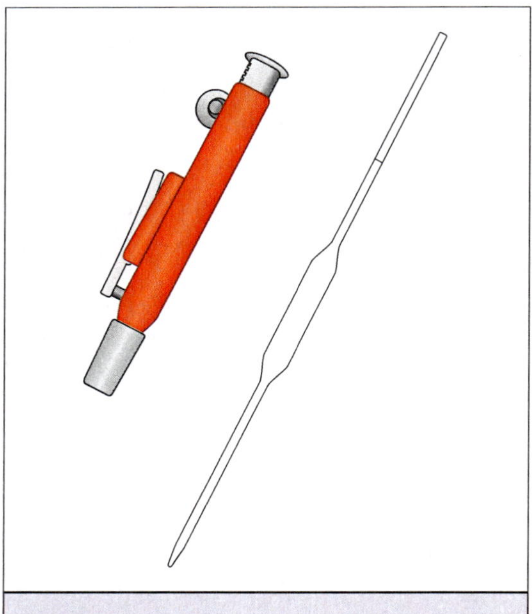

**FIGURE 5.1**

12. Put the tip of the pipette into the unknown acetic acid, and use the thumbwheel to pull up *exactly* 10.00 mL of the acetic acid. If you pull up too much or too little, use the thumbwheel to adjust until the meniscus is *exactly* at the 10.00-mark.

13. Put the pipette tip into the Ernlenmeyer flask ,and completely discharge the contents of the pipette by rolling the thumbwheel down.

14. Add 2–3 drops of phenolphthalein solution to the flask.

15. Put a stirring bar into the flask. Slide the bar down the side so that it won't break the bottom out of the flask.

16. Put the flask on the stir station below the burette tip. Put a piece of white filter paper under the flask so the pink color will be easier to see when it appears. Begin stirring the solution.

**FIGURE 5.2**

# Part III: Reaction of the NaOH with the Acetic Acid

As sodium hydroxide reacts with the acetic acid, you will begin to see traces of a pink color. This comes from the phenolphthalein indicator that you added. When you have reached the point where you have added just enough NaOH to react with the acetic acid, the solution will be a ***very pale*** pink color.

Your first reaction will be a trial run that will give you a rough idea of how much NaOH it will take to react with the unknown acetic acid. You can perform this first reaction rather quickly without taking a lot of time to add small incre-

**FIGURE 5.3**

ments of NaOH. After you know about how much NaOH it takes for the reaction, you will then do three more repeats of the reaction. But you will do these more slowly near the end so that you can add just enough NaOH to make the final solution just barely pink.

## Trial Run

17. The flask containing the measured volume of unknown acetic acid should be set up on the stirring stand under the burette. *Make sure you have recorded the initial volume of NaOH.*

18. Open the stopcock, and let the NaOH flow at a slow but steady rate into the flask. Watch carefully for the pink color to appear. Eventually the pink color will stay in the solution. Close the stopcock when the solution is pink. Record the final volume. Now you know *approximately* how much NaOH it will take to react with 10 mL of unknown acetic acid

19. Pour the contents of the flask into the large beaker you are using to collect reaction mixes in. Rinse the flask well with water. You don't need to dry the flask. Don't let the stirring bar doesn't go into the beaker.

## Accurate Runs

20. Add 10.00 mL of unknown acetic acid to the flask using the volumetric pipette just as you did before. Add 2–3 drops of phenolphthalein. Put the flask under the burette tip. Stir.

21. Add enough additional NaOH to the burette to re-fill it. Record the initial volume.

22. Now begin adding NaOH to the unknown acetic acid. You can add it quickly at first. But, when you get to within about 2 mL of the approximate volume you found in the trial run, turn off the stopcock.

23. Now you will add the NaOH **a little at a time** until the solution in the flask has *just barely turned pink* and stayed pink. Try adding just a drop or two of NaOH at a time, close the stopcock, and let the reaction mixture stir for a moment. When you think the solution is the right color (*a very pale pink color that stays in the solution*), stop adding the NaOH. Record the final volume of NaOH.

24. Discard the reaction mixture into the large beaker.

25. Repeat Steps 20–23 two more times with fresh 10.00 mL volumes of unknown acetic acid.

## Clean-Up and Disposal

- Rinse the Erlenmeyer flask well after you have disposed of the final reaction mixture. Return it to the glassware drawer.

- Return the rinsed stirring bar to the drawer with the flask.

- Empty any remaining NaOH from the burette into the large beaker. Rinse the burette well with distilled water. Return the burette to the stirring stand.

- Test the pH of the material in the large beaker. Notice that pH paper is **not** the same as litmus paper. If it is between 5 and 9, you can pour the contents down the drain. If it is above or below this pH, ask your instructor for assistance.

# DATA SHEET

Name: _________________________________     Grade: _________________________________

Date Experiment Performed: _______________     Days Late: _____________________________

CRN of Lab Section: _____________________     Instructor's Initials: ___________________

| General Grading Items | 20 Points |
|---|---|
| Copies of Lab Pages Submitted; Labeled with Name and Date, Complete Information, Readable, Data Recorded Matches Results Given in Report | |
| Name and CRN Included on Report | |
| All Safety Rules Were Followed | |
| Waste Was Properly Disposed of and Lab Area Was Cleaned | |
| Evaluation of Student Performance Overall (Student Was on Time, Followed Safety Rules, Performed the Lab Correctly and Within the Time Allowed, Etc.) | |
| 20-Question Pre-Lab Assignment | /20 |
| **Total** | **/20** |

**CHE 111L student: Submit the copies of your lab notebook pages for this experiment to your TA before you leave lab for the day.**

# Questions and Calculations (80 points)

**1.** Write the balanced chemical equation for the reaction between sodium hydroxide and acetic acid. (5 points)

**2.** What was the concentration, in moles/liter, of the NaOH you used for this reaction?

_______________ M (2 points)

**3.** Perform the calculations in the table below for each of the three runs of the reaction. ***You don't need to include the trial run.*** (40 points)

***Show the complete calculation for Trial 1 in the provided spaces in Table 5.2. You can show results only in the table itself.***

| TABLE 5.2 | | | |
|---|---|---|---|
| | **Trial 1 Results** | **Trial 2 Results** | **Trial 3 Results** |
| Final Volume NaOH mL | | | |
| Initial Volume NaOH mL | | | |
| Volume NaOH Used, mL | | | |
| Mole Ratio between NaOH and Acetic Acid (from the balanced chemical equation) | | | |
| Moles of NaOH That Reacted with the Acetic Acid in the Vinegar (see below) | | | |
| Show this calculation for Trial 1 data only here (remember that M = moles/liter). | | | |

**TABLE 5.2**

| | Trial 1 Results | Trial 2 Results | Trial 3 Results |
|---|---|---|---|
| Moles of Acetic Acid Present in the Unknown (see below) | | | |

Show this calculation for Trial 1 data only here (look at the equation: what's the relationship between moles of NaOH and moles of acetic acid?).

| | Trial 1 Results | Trial 2 Results | Trial 3 Results |
|---|---|---|---|
| Grams of Acetic Acid Present in the Unknown Vinegar (see below) | | | |

Show this calculation for Trial 1 data only here.

| | Trial 1 Results | Trial 2 Results | Trial 3 Results |
|---|---|---|---|
| Volume of Unknown Acetic Acid Used, mL | | | |
| % (w/v) of Acetic Acid in the Vinegar (see below) | | | |

Show this calculation for Trial 1 data only here.

**4.** Calculate the mean and standard deviation for the three trials you did to determine the % (w/v) of acetic acid in the vinegar. (5 points)

  **a.** Mean:                         **b.** Standard deviation:

**5.** The unknown was actually commercial vinegar, which is 5% acetic acid. Calculate the % error of your result. (5 points)

6.  Your stomach fluids have a pH of about 1–2 and would turn litmus paper a red color. Is your stomach fluid acidic or basic? (4 points)

7.  Milk of Magnesia is a product known as an *antacid*. What is the chemical compound that is the active ingredient of Milk of Magnesia? (If you don't know, look it up.) (5 points)

8.  A drop of Milk of Magnesia would turn litmus paper blue. Is this an acid or a base? (4 points)

9.  What effect does the Milk of Magnesia have on the stomach fluids? (5 points)

10. Write a balanced chemical equation that shows the reaction of stomach acid (hydrochloric acid) with the active ingredient of Milk of Magnesia. (5 points)

# Oxidation–Reduction Reactions



## Objectives

At the completion of the lab, you should be able to

- define reduction, oxidation, reducing agent, oxidizing agent, reduction product, and oxidation product;

- assign oxidation numbers correctly to each element in a redox reaction;

- determine which species in a redox reaction

    - is reduced

    - is oxidized

    - acts as the reducing agent

    - acts as the oxidizing agent

    - is the reduction product

    - is the oxidation product; and

- list species according to their relative strengths as reducing agent or oxidizing agents.

# 20-QUESTION PRE-LAB ASSIGNMENT

CRN:________________________________     Your Name: ______________________________

Date Submitted: ____________________     TA's Name: _______________________________

*Please write your answers legibly in the space provided. Many answers will actually consist of three or four pieces of information that are clearly written on the board during the conversation. You must include all aspects of the answer to receive full credit.*

**1.** Which two words is redox a contraction of?

**2.** Define oxidation.

**3.** Define reduction.

**4.** Define oxidizing agent.

**5.** Define reducing agent.

**6.** List the five classes of chemical reactions.

**7.** Which of the above five classes are **always** redox reactions?

**8.** Which of the five classes are **never** redox reactions?

**9.** Which of the five classes sometimes **is** and sometimes is **not** redox?

**10.** Compare and contrast hydrogen replacement vs. metal replacement vs. anion replacement.

For the questions below, complete the sentences by circling the correct choice in parenthesis.

**11–12.** Metals are usually (oxidized/reduced) and so are (reducing agents/oxidizing agents).

**13–14.** Metal cations are usually (oxidized/reduced) and so are (reducing agents/oxidizing agents).

**15–16.** The Halogens are usually (oxidized/reduced) and so are (reducing agents/oxidizing agents).

**17–18.** Halide ions are usually (oxidized/reduced) and so are (reducing agents/oxidizing agents).

**19–20.** For the redox process below, assign charges to every pertinent species, decide which atom is oxidized, which atom is reduced, as well as the reducing agent and the oxidizing agent.

$$Cu(s) + Au(NO_3)_3(aq) \rightarrow Au(s) + Cu(NO_3)_2(aq) \quad \text{[not balanced]}$$

# BACKGROUND

In oxidation-reduction reactions, electrons are exchanged between two species. These are very common types of reactions. Oxidation and reduction will always occur together since one species will lose electrons and another species will gain them.

Some basic definitions are needed for understanding this experiment.

**Oxidation:** the loss of electrons by a species, resulting in an *increase* in oxidation number. (This can also mean the addition of oxygen to a species.)

**Reduction:** the gain of electrons by a species, resulting in a *decrease* in oxidation number. (This can also mean the addition of hydrogen to or removal of oxygen from a species.)

**Reducing agent (or reductant):** this is the substance that reduces something else. It is the species that gives the electrons to the substance being reduced. So, the reducing agent is oxidized.

**Oxidizing agent (or oxidant):** this is the substance that oxidizes something else. It is the species that gets the electrons lost by the substance being reduced. So, the oxidizing agent is actually reduced during the reaction.

Be sure you understand how these terms are applied in the example below:

$$Fe^{2+} \quad + \quad Co^{3+} \quad \rightarrow \quad Fe^{3+} \quad + \quad Co^{2+}$$

| $Fe^{2+}$ | $Co^{3+}$ | $Fe^{3+}$ | $Co^{2+}$ |
|---|---|---|---|
| Is Oxidized<br>Is the Reducing<br>Agent | Is Reduced<br>Is the Oxidizing<br>Agent | Oxidation Product<br>Lost 1 Electron<br>to $Co^{3+}$ | Reduction Product<br>Gained 1 Electron<br>from $Fe^{2+}$ |

## Recognizing a Redox Reaction

In order to recognize a redox reaction, you need to understand the concept of oxidation numbers and be able to assign oxidation numbers correctly. Oxidation–reduction reactions are covered in your textbook on pages 174–188 of Silberberg and Amateis, 8e.

## Materials

- small test tubes
- test tube rack
- test tube holder
- neutral litmus paper
- abrasive paper (for cleaning surfaces of metal sample)
- 125-mL Erlenmeyer flask with stopper
- 100-mL graduated cylinder

## Solutions and Reagents

- calcium metal pellets
- magnesium strips
- iron nails
- copper wire
- zinc strips
- chlorine water, $Cl_2(aq)$
- bromine water, $Br_2(aq)$
- iodine water, $I_2(aq)$

- 1 M HCl
- 6 M HCl
- 0.02 M $ZnSO_4$
- 0.02 M $MgSO_4$
- 0.02 M $CuSO_4$
- $FeSO_4 \cdot 7\,H_2O$ (solid)
- 0.1 M NaCl
- 0.1 M NaBr
- 0.1 M NaI

# PROCEDURE

You will run all of the following reactions. For each reaction, use the dispenser to get the required reagent volumes or a very small piece of the solid material if the reagent is a solid. Return to your lab station with the reagents. In your lab notebook, describe observations and results for each experiment. Do not simply state that a reaction has occurred—describe what it looked like. If a reaction has occurred, tell why you think so based on your observations and results.

<table>
<tr><td>

**SAFETY!**

**WASTE DISPOSAL**

</td><td>

After each reaction, decant the liquid material into the waste container provided for the solutions leaving any unreacted metal in the test tube. Then rinse with water the unreacted metal sample and place it onto a paper towel to dry *or* into the appropriate collection container so that it may be recycled and reused.

</td></tr>
</table>

## Part I: The Reducing Strengths of Metals

Metals most often act as reducing agents since they give up their electrons relatively easily. In this part of this experiment, you will compare the reducing strengths of several metals.

### Reducing Strength of Metals with Water

1. **Reaction of Na and K with Water**

   There is a video clip in Blackboard that shows the reaction of Na and K with water. Watch the video and record your observations in your lab notebook.

2. **Reaction of Mg with Water**

   Get one small piece of Mg metal and 1 mL of water in separate test tubes, and take them back to your lab bench. Before running the reaction, make the following observations and record them *in your lab notebook.*

   a. Describe the appearance of the Mg metal before reaction.

   b. Test the pH of the water on a piece of litmus paper.

   c. Put the tube containing Mg into a rack. Add the water. Record your observations of what happens after mixing the water and Mg metal in your lab notebook.

   d. Test the pH of the solution in the tube after the reaction.

### 3. Reaction of Ca with Water

Get one small piece of Ca metal and 1 mL of water in separate test tubes, and take them back to your lab bench. Before running the reaction, make the following observations and record them *in your lab notebook.*

**a.** Describe the appearance of the Ca metal before reaction.

**b.** Test the pH of the water on a piece of litmus paper.

**c.** Put the tube containing Ca into a rack. Add the water. Record your observations of what happens after mixing the water and Ca metal in your lab notebook.

**d.** Test the pH of the solution in the tube after the reaction.

## Reducing Strength of Metals (Mg, Fe, Cu, Zn) with Hydrochloric Acid (HCl)

### 4. Reaction of Mg with HCl

Get one small piece of Mg metal and 1 mL of 1 M HCl in separate test tubes, and take them back to your lab bench. Before running the reaction, make the following observations and record them *in your lab notebook.*

**a.** Describe the appearance of the Mg metal before reaction.

**b.** Test the pH of the HCl on a piece of litmus paper.

**c.** Put the tube containing Mg into a rack. Add the HCl. Record your observations of what happens after mixing the water and Mg metal in your lab notebook.

**d.** Test the pH of the solution in the tube after reaction.

> ***If there has been no reaction:***
>
> **i.** If there appears to be no reaction, heat the tube *gently* over a small burner flame. Do not let the liquid boil out of the tube! Note the evidence for a reaction, if any.
>
> **ii.** If there is still no reaction with heat, decant the 1 M HCl into a waste container. Leave the metal in the tube. Add 6 M HCl. Let the tube stand and see if there is a reaction.
>
> **iii.** If there is still no reaction, heat the tube *gently* over a small burner flame. Do not let the liquid boil out of the tube! Note the evidence for a reaction, if any.

**Acids being heated! Use caution!**

### 5. Repeat Step 4 with Fe, Cu, and Zn. Record observations as above.

<table>
<tr><td>SAFETY!<br><br>WASTE DISPOSAL</td><td>After each reaction, decant the liquid material into the waste container provided for the solutions leaving any unreacted metal in the test tube. Then rinse the unreacted metal sample with water and lay it out to dry on the paper toweling so that it may be recycled.</td></tr>
</table>

# Part II: The Reducing Strengths of Metals Compared to Each Other

In this part of the experiment you will compare the ability of various metals to oxidize or reduce each other.

**6. Mg vs. Zn**

   **a.** In a clean test tube place a strip of freshly abraded Mg into about 2 mL of 0.02 M $ZnSO_4$. Leave a section of the Mg strip sticking up above the liquid. This will make it easier to see if any reaction has taken place.

   **b.** Similarly, place a strip of freshly abraded Zn into about 2 mL of 0.02 M $MgSO_4$. Leave a section of the Zn strip sticking up above the liquid. This will make it easier to see if any reaction has taken place.

   **c.** Allow each system to stand until a change is clearly detectable in at least one system. This may require a few minutes. Be patient! Record your observations in your lab notebook.

<table>
<tr><td>SAFETY!<br><br>WASTE DISPOSAL</td><td>After each reaction, decant the liquid material into the waste container provided for the solutions leaving any unreacted metal in the test tube. Then rinse the unreacted metal sample with water and place it onto the paper toweling to dry before recovery.</td></tr>
</table>

**7. Cu vs. Zn**

   **a.** In a clean test tube place a strip of freshly abraded Cu into about 2 mL of 0.02 M $ZnSO_4$. Leave a section of the Cu strip sticking up above the liquid. This will make it easier to see if any reaction has taken place.

   **b.** Similarly, place a strip of freshly abraded Zn into about 2 mL of 0.02 M $CuSO_4$. Leave a section of the Zn strip sticking up above the liquid. This will make it easier to see if any reaction has taken place.

   **c.** Allow each system to stand until a change is clearly detectable in at least one system. This may require a few minutes. Be patient! Record your observations in your lab notebook.

**8. Fe vs. Cu**

   **a.** $FeSO_4$ must be freshly made since the $Fe^{2+}$ in the solution will quickly react with the oxygen in the air. Get a 125-mL Erlenmeyer flask and stopper. Add about 0.3 g of $FeSO_4$ to the flask and add about 50 mL of distilled water. Swirl until the solid is dissolved. Keep the solution stoppered when you are not using it.

   **b.** In a clean test tube place a strip of freshly abraded Fe into about 2 mL of 0.02 M $CuSO_4$. Leave a section of the Fe strip sticking up above the liquid. This will make it easier to see if any reaction has taken place.

   **c.** Similarly, place a strip of freshly abraded Cu into about 2 mL of the freshly-made $FeSO_4$ solution. Leave a section of the Cu strip sticking up above the liquid. This will make it easier to see if any reaction has taken place.

   **d.** Allow each system to stand until a change is clearly detectable in at least one system. This may require a few minutes. Be patient! Record your observations in your lab notebook.

**9. Fe vs. Zn**

   Figure out how to set up the metals and solutions to compare the relative reducing strengths of Fe and Zn. Use the previous sections as an example.

# Part III: The Reducing Strength of Halide Ions and Oxidizing Strengths of Elemental Halogens

In this section, you will compare the reducing strengths of $Cl^-$, $Br^-$, and $I^-$.

The work in this section requires that all solutions used are ***carefully examined for color before and after reactions.*** This is because a *change* in color indicates reaction, while color dilution (reduction in color intensity) does not. For instance, if a small amount of grape Kool-Aid is added to a large amount of fresh water, the resulting dilute Kool-Aid is still purple in color although the intensity of the purple color in the dilute Kool-Aid is much less than in the original. This is still a solution of Kool-Aid, and the reduction in color intensity has nothing whatsoever to do with any reaction.

**10. $Cl^-$ with $Br_2$**

   **a.** Get 1 mL of 0.1 M NaCl and 1 mL of bromine water ($Br_2$ dissolved in water) in separate test tubes. Record the color of each solution.

   **b.** Mix the solutions. Watch for any change in color of the solution.

**11. $Br^-$ with $Cl_2$**

   **a.** Get 1 mL of 0.1 M NaBr and 1 mL of chlorine water ($Cl_2$ dissolved in water) in separate test tubes. Record the color of each solution.

   **b.** Mix the solutions. Watch for any change in color of the solution.

**12. Br$^-$ and I$^-$ with I$_2$ and Br$_2$, Respectively**

Design and perform the experiment necessary to determine the relative reducing strengths of Br$^-$(aq) and I$^-$(aq), and the relative oxidizing strengths of Br$_2$ and I$_2$. Use Steps 10–11 as an example. Record your observations in your lab notebook.

**13. I$^-$ and Cl$^-$ with Cl$_2$ and I$_2$, Respectively**

Design and perform the experiment necessary to determine the relative reducing strengths of Cl$^-$(aq) and I$^-$(aq), and the relative oxidizing strengths of Cl$_2$ and I$_2$. Use Steps 10–11 as an example. Record your observations in your lab notebook.

# Clean-Up and Disposal

- Dispose of all solutions in the correct waste container. Rinse the metal pieces with water and put them on the paper toweling to dry.
- Rinse all test tubes well with water, and return them to the test tube rack.
- Put all equipment back into the lab drawer.
- Turn in the carbon copy of your data to your TA before you leave.

# DATA SHEET

Name: _______________________________     Grade: _______________________________

Date Experiment Performed: _______________     Days Late: _______________________________

CRN of Lab Section: _______________________     Instructor's Initials: _______________________

| General Grading Items | 20 Points |
|---|---|
| Copies of Lab Pages Submitted; Labeled with Name and Date, Complete Information, Readable, Data Recorded Matches Results Given in Report | |
| Name and CRN Included on Report | |
| All Safety Rules Were Followed | |
| Waste Was Properly Disposed of and Lab Area Was Cleaned | |
| Evaluation of Student Performance Overall (Student Was on Time, Followed Safety Rules, Performed the Lab Correctly and Within the Time Allowed, Etc.) | |
| 20-Question Pre-Lab Assignment | /20 |
| **Total** | /20 |

**CHE 111L student: Please submit a copy of your lab notebook pages with today's data to your TA before you leave.**

Review the material in Chapter 4 of your text (Silberberg and Amateis, 8e, pages 174–188) for additional explanation of these topics:

- Oxidation and reduction processes
- Determining oxidation number for an element
- Producing tables of redox strength (activity series)

# General Questions on Oxidation–Reduction Processes and Terms (20 points)

Answer the questions for the reaction below:

$$MnO_4^-(aq) + Br^-(aq) \rightarrow Br_2(l) + Mn^{2+}(aq)$$

(The reaction is not balanced, but you don't need that to answer the questions below.)

1. Determine the oxidation number of each element in this reaction

   **a.** Mn (in the $MnO_4^-(aq)$):

   **b.** O (in the $MnO_4^-(aq)$):

   **c.** $Br^-(aq)$:

   **d.** $Br_2(l)$:

   **e.** $Mn^{2+}(aq)$:

2. Identify the following species in the reaction above:

   **a.** What is oxidized?

   **b.** What is reduced?

   **c.** What is the oxidation product?

   **d.** What is the reduction product?

   **e.** What is the oxidizing agent?

   **f.** What is the reducing agent?

3. Suppose you run a series of reactions involving the conjugate reduction/oxidation pairs of Co, Al, and Au. You see the following results:

   - Co metal in $Al^{3+}$ solution: no reaction
   - Al metal in $Co^{2+}$ solution: reaction
   - Au metal in $Co^{2+}$ solution: no reaction
   - Co metal in $Au^{3+}$ solution: reaction
   - Al metal in $Au^{3+}$ solution: reaction
   - Au metal in $Al^{3+}$ solution: no reaction

   List the three pairs in the correct order in the table. Put the reduced form of the pair in the left column, and the oxidized form in the right column.

| TABLE 6.1 | |
|---|---|
| **Strongest Reducing Agent** | **Weakest Oxidizing Agent** |
| | |
| | |
| | |
| **Weakest Reducing Agent** | **Strongest Oxidizing Agent** |

4. Is the reaction below an oxidation-reduction reaction? How do you know?

   $$AgNO_3 + NaCl \rightarrow AgCl(s) + NaNO_3$$

# Part I: Reducing Strength of Metals (15 points)

## Reaction of Na, K, Mg, Ca with Water

**TABLE 6.2**

| Summary of Results | Relative Vigor of the Reaction Compared to Other Metals in This Part (1 = most reactive; 4 = least reactive) | Litmus Paper Test Results after Reaction Has Taken Place |
| --- | --- | --- |
| Reaction of Na with Water | | |
| Reaction of K with Water | | |
| Reaction of Mg with Water | | |
| Reaction of Ca with Water | | |

5. Based on your observations of the litmus paper test, which species was present in highest concentration after each of the reactions: $OH^-$ or $H^+$? How do you know this?

6. Look at the elements in the first column of the periodic table, and locate the elements Na and K. ***Based on your observations,*** would you predict that rubidium would react more or less strongly with water than sodium did? Explain your prediction.

7. Look at the elements in the second column of the periodic table, and locate Ca and Mg. ***Based on your observations,*** would you predict that Sr would react more or less strongly with water than Mg did? Explain your prediction.

**8.** Write the complete ionic equation for the reaction of K with water.

Identify the reducing and oxidizing agents and the oxidation and reduction products.

**a.** Oxidizing agent:

**b.** Reducing agent:

**c.** Oxidation product:

**d.** Reduction product:

## Reducing Strength of Mg, Fe, Cu and Zn with Hydrochloric Acid (15 points)

**9.** Summarize the results of your reactions.

**TABLE 6.3**

| | Relative Vigor of the Reaction with No Heating (1 strongest reaction to 4 weakest or no reaction) | Relative Vigor of the Reaction with Heating | Gas Formed |
|---|---|---|---|
| Reaction of Mg with HCl | | | |
| Reaction of Fe with HCl | | | |
| Reaction of Cu with HCl | | | |
| Reaction of Zn with HCl | | | |

**10.** Make a table like the one in Question 4 to show the order of Cu, Fe, Mg, and Zn from strongest to weakest reducing agent.

Put the ions of the metals in order from weakest to strongest oxidizing agent.

**TABLE 6.4**

| Strongest Reducing Agent | Weakest Oxidizing Agent |
|---|---|
| | |
| | |
| | |
| | |
| **Weakest Reducing Agent** | **Strongest Oxidizing Agent** |

**11.** Write the full ionic equation for the reaction of Mg with HCl. Identify the reducing and oxidizing agents and the reduction and oxidation products.

**a.** Reducing agent:

**b.** Oxidizing agent:

**c.** Reduction product:

**d.** Oxidation product:

# Part II: The Oxidizing Strength of Metal Ions (15 points)

**12.** Indicate whether or not a reaction occurred for the following mixtures you made in lab.

**a.** Mg in $ZnSO_4$:

**e.** Fe in $CuSO_4$:

**b.** Zn in $MgSO_4$:

**f.** Cu in $FeSO_4$:

**c.** Cu in $ZnSO_4$:

**g.** Fe in $ZnSO_4$:

**d.** Zn in $CuSO_4$:

**h.** Zn in $FeSO_4$:

**13.** For every mixture above where you said there was a reaction, write the net ionic equation for the reaction.

**14.** Make a table like the one in Question 4 to show the order of Cu, Fe, Mg, and Zn from strongest to weakest reducing agent.

Put the metal ions in order from weakest to strongest oxidizing agent.

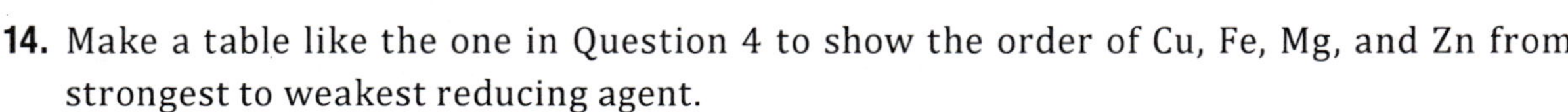

| TABLE 6.5 | |
|---|---|
| **Strongest Reducing Agent** | **Weakest Oxidizing Agent** |
| | |
| | |
| | |
| | |
| **Weakest Reducing Agent** | **Strongest Oxidizing Agent** |

**15.** Compare the table you just made with the table in Question 10, Part B. Are they the same or different?

# Part III: The Reducing Strength of Halide Ions and Oxidizing Strengths of Elemental Halogens (15 points)

**16.** Indicate whether or not a reaction occurred for the following mixtures you made in lab.

**a.** $Cl^-$ in $Br_2$:

**b.** $Br^-$ in $Cl_2$:

**c.** $Br^-$ in $I_2$:

**d.** $I^-$ in $Br_2$:

**e.** $Cl^-$ in $I_2$:

**f.** $I^-$ in $Cl_2$:

**17.** For every mixture above where you said there was a reaction, write the net ionic equation for the reaction.

**18.** Make a table like the one in Question 4 to show the order of these agents from strongest to weakest reducing and oxidizing agent.

| TABLE 6.6 | |
|---|---|
| **Strongest Reducing Agent** | **Weakest Oxidizing Agent** |
| | |
| | |
| | |
| **Weakest Reducing Agent** | **Strongest Oxidizing Agent** |

# Gas Laws



## Objectives

At the completion of the lab, you should be able to

- use Logger Pro to produce a properly labeled graph;

- use Logger Pro to get a linear fit of a data set;

- give the equation for a straight-line and state the meaning of the intercept and slope;

- describe the physical meaning of the slope and intercept of a plot of density data of a mixture;

- state the difference between a direct and an inverse relationship between two physical quantities;

- state Boyle's law, Charles' law, Avogadro's hypothesis, and the ideal gas law;

- interpret plots of gas pressure, temperature, and/or volume and provide the physical meaning of the slope and intercept;

- obtain experimental data for the pressure and temperature of a known mass of gas and made graphs of the data using Logger Pro; and

- interpret the plots to obtain values for other gas parameters.

# 20-QUESTION PRE-LAB ASSIGNMENT

CRN:_________________________________________     Your Name: ___________________________________

Date Submitted: ______________________________     TA's Name: ___________________________________

*Please write your answers legibly in the space provided. Many answers will actually consist of three or four pieces of information that are clearly written on the board during the conversation. You must include all aspects of the answer to receive full credit. Happy YouTubing!*

**1.** Show the progression of a scientific idea from a hypothesis through a law, and define each step.

**2.** Write the definition of a law.

**3.** Provide the non-chemistry example of a law.

**4.** What are the four variables that must be specified to completely describe gas behavior?

**5.** Which units are used when we make these measurements in lab?

**6.** Which units are these variable often converted to?

**7.** List the three gas laws we work with this week in lab, and write out the relationships between the variables.

**8.** For plots of $V$ vs. $T$ and $P$ vs. $T$, which quantity is plotted on the $x$-axis?

**9.** Write out the ideal gas law and show how these variables result in a straight-line plot.

**10.** Sketch and label a plot of the ideal gas law.

**11.** Does air have mass? Explain your answer.

**12.** Write the percent composition of air, and set up the calculation to obtain the molar mass of air.

**13.** Oops! Dr. May wrote an incorrect value for $R$ on the board—it was close, but missing a digit. What was the wrong value for $R$ that he wrote, and what is the correct value you should use in your % error calculation?

**14.** What is the tolerance (or the range) within which % error is considered to be accurate?

**15.** How cold is absolute zero?

**16.** What is that actual numerical temperature in °C, °F, and kelvins?

**17.** What is the definition of **absolute zero?**

**18.** According to the kinetic molecular theory of gases, what is temperature directly proportional to? And how does that translate into absolute zero?

**19.** What happens to the pressure of a gas at absolute zero?

**20.** For the $P$ vs. $T$ plot, what is **not** useful and what **is** useful about the straight-line plot?

# BACKGROUND

Gases are one of the three states of matter. You will do this experiment dealing with gases in two parts.

- In Part I, you will review the gas laws and use Logger Pro to plot data from an ideal gas. You will also interpret the physical meaning of the graphs.

- In Part II, you will take actual empirical measurements of pressure and temperature for a gas.

## Materials

The equipment needed for this experiment should be in a plastic box on the desk top, except for the ring stand and clamp. Those two items should be left out on the desk top.

- Vernier unit
- pressure probe
- temperature probe
- syringe
- 125-mL flask with 2-hole stopper
- 600-mL beaker
- connectors
- ring stand
- clamp

# PROCEDURE

## Part I: Exploring the Properties of Gases and Performing a Graphical Analysis of Ideal Gas Data Using Logger Pro

### Brief Review of Gases and Gas Laws

A gas has no definite shape or volume. The volume of gas is much more affected by pressure and temperature than are the volumes of either solids or liquids. So, we need to understand these quantities and how they are related to each other in order to describe gases.

The **temperature** of gases is normally expressed in the Kelvin scale, or the absolute temperature scale. It is simple to convert from °C to K: simply add 273.15—or just 273 in many cases. The temperature in kelvins is always greater than zero; there are no negative kelvin temperatures. The **pressure** of a gas is most often expressed in either atmospheres (atm) or in mm of mercury (mm Hg, or just mm). **Volume** of a gas is often given in liters.

There are three gas laws that you can learn more about in Chapter 5 of Silberberg and Amateis. These include Boyle's law, Charles' law, and Avogadro's hypothesis. These laws relate the number of moles of gas ($n$), the pressure ($P$), the temperature ($T$) and the volume ($V$).

*Boyle's law:* Suppose you have a fixed amount of gas at a constant temperature, like a syringe full of air. You know that if you increase the pressure on the gas by pushing down on the syringe handle, you can reduce the volume of the gas. This illustrates Boyle's law. It states that, for constant temperature and moles of gas, the volume of a gas will vary inversely with pressure.

$$V \approx \frac{1}{P} \quad \text{or} \quad V = \frac{k}{P}$$

where $k$ is a constant.

*Charles' law:* Suppose you have a fixed amount of gas at a constant pressure, like a balloon full of air. You know that if you increased the temperature of the gas in the balloon by putting it into a bowl of hot water, the balloon would increase in volume. Or, if you decreased the temperature of the gas in the balloon by putting it into a bowl of cold water, the balloon would decrease in volume. This illustrates Charles' law. It states that, for constant pressure and moles of gas, the volume of a gas will vary directly with temperature.

$$V \approx T \quad \text{or} \quad V = kT$$

where $k$ is a constant.

*Avogadro's hypothesis:* A third important law says that the same number of molecules of gas will occupy the same volume at the same temperature and pressure. The identity of the gas doesn't matter—only the number of gas molecules.

$$V \approx n \quad \text{or} \quad V = kn$$

where $k$ is a constant.

***Ideal gas law:*** If you put these three observations together, you get the *ideal gas law,* which is

**$PV = nRT$**

where $R$ is the ideal gas law constant. It has a value of 0.0821 L·atm/mol·K.

## Plotting a Graph of Pressure vs. Temperature Data

**1.** The data in Table 7.1 was collected for 0.125 moles of nitrogen gas contained in a rigid 3.00 liter vessel.

**TABLE 7.1**

| Pressure (atm) | Temperature (°C) |
|---|---|
| 1.022 | 26.9 |
| 1.01 | 21.5 |
| 0.995 | 17.6 |
| 0.975 | 11.6 |
| 0.96 | 6.6 |
| 0.939 | 2.6 |
| 0.932 | −0.9 |

   **a.** Use Logger Pro to construct a properly labeled and scaled graph of pressure ($y$-axis) vs. temperature ($x$-axis, in degrees C). Uncheck "connect points" in "graph options." Do a linear fit of the data, be sure the slope and intercept information is visible, and make a hard copy of the graph. *(You will need to attach this graph to your Data Sheet.)*

   **b.** Use the graph to determine the value of the ideal gas constant in units of L·atm/mol·K. Show your calculations on your Data Sheet.

   **c.** Use the graph to determine the value of absolute zero in units of degrees C. Record your answer on the graph, and briefly explain how you arrived at this value.

## Plotting a Graph of Pressure vs. Volume Data

**2.** The data in Table 7.2 was collected for 0.750 moles of argon gas at 25°C.

**TABLE 7.2**

| $P$ (atm) | $V$ (L) |
|---|---|
| 0.60 | 30.6 |
| 0.75 | 24.4 |
| 1.01 | 18.1 |
| 1.48 | 12.4 |
| 2.02 | 9.06 |
| 2.50 | 7.32 |
| 3.03 | 6.04 |
| 4.04 | 4.53 |

   **a.** Use Logger Pro to construct a properly labeled and scaled graph of volume ($y$-axis) vs. pressure ($x$-axis). Make a hard copy of the graph. *(You will need to attach this graph to your Data Sheet.)*

   **b.** Is this an inverse or direct relationship?

   **c.** Is this a linear relationship?

**d.** In Logger Pro, click "analyze" from the top menu bar, then select "curve fit." Click on "linear fit," and then click "try fit" button at the bottom. You'll see an attempt at fitting a straight-line to the *P* vs. *V* data. Obviously, a straight-line does not fit this data very well.

**e.** Try some of the other types of equations. Which one do you find that fits this set of data the best? Record the name of the fit and the equation in your notebook.

**f.** Record the value for the constant (*A*) in the fit you selected for your graph.

## Part II: Empirical Measurements of the Pressure and Temperature of a Gas

In this part of the experiment, you will obtain your own experimental data, plot the data, and use your own data to answer questions about the gas laws.

**3.** The connectors should already be in place in the stopper. Notice that the stopcock is **open** when the handle is vertical; it is **closed** when the handle is perpendicular. See Figure 7.1 and Figure 7.2. Insert the stopper into the flask.

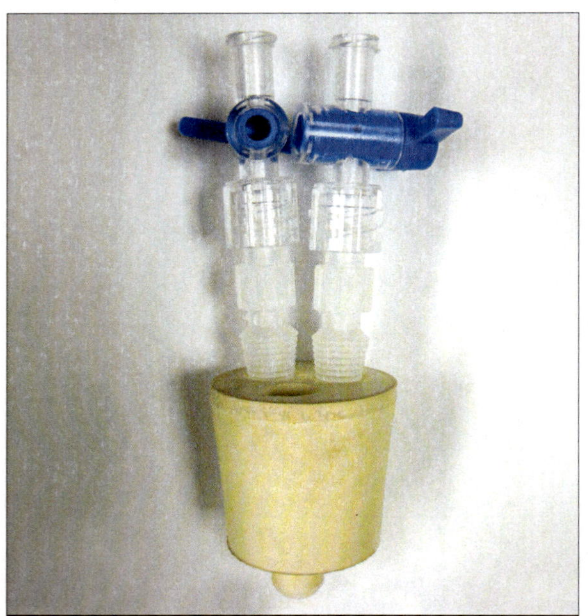

**FIGURE 7.1 Connectors in Stopper** This unit should already be assembled and be in the supplies box.

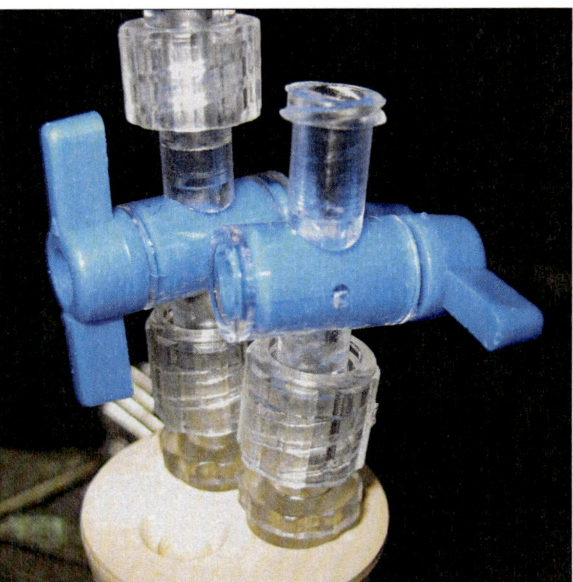

**FIGURE 7.2 Stopcock** The blue stopcock in the vertical position is **open** (left). The blue stopcock in the horizontal position is **closed** (right).

4. First, you need to remove most of the air from the flask. Close both stopcocks. Attach the syringe to one of the connectors (see Figure 7.3). The syringe and connector are threaded, so they are screwed together. Be sure the other stopcock remains closed. Open the stopcock that is attached to the syringe, and then pull out the syringe plunger to draw air out of the flask.

   ***Close the stopcock,*** then unscrew the syringe from the connector and expel the air. Reconnect the syringe, ***open the stopcock,*** and withdraw more air. Repeat this about 8–10 times.

   ***Don't forget to close the stopcock before removing the syringe!***

5. Now weigh the empty flask with stopper and two stopcocks. Read to three decimal places. *Record all data directly into the Data Sheet.*

6. Now you will let the flask fill with air, and inject extra air into it. Open one of the stopcocks, and let air go into the flask. Get the syringe, and pull the plunger out as far as it will go without coming out. Attach the syringe to the open stopcock, and inject a syringeful of air into the flask. **Close the stopcock.** Remove the syringe, raise the plunger and re-attach the syringe. Open the stopcock, and inject another syringeful of air into the flask.

7. ***Be sure the stopcock is closed,*** and then remove the syringe.

8. Re-weigh the flask with the stopper and two stopcocks. Record the mass on the Data Sheet. Determine the mass of air in the flask.

9. Clamp the flask at the neck, and lower it into the 600-mL beaker.

10. Get the Vernier unit, a pressure probe, and a temperature probe. Connect the gas pressure probe to Channel 1 of the Vernier unit. Connect the temperature probe to Channel 2 (see Figure 7.4).

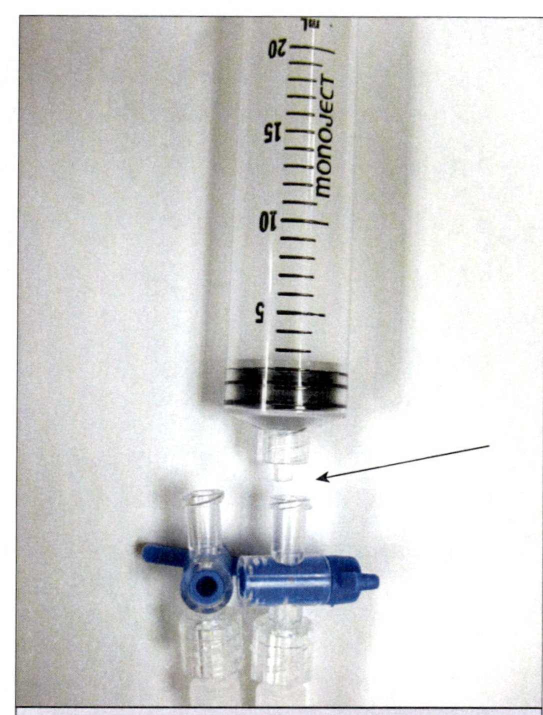

**FIGURE 7.3   Screw Syringe Gently into Connector**

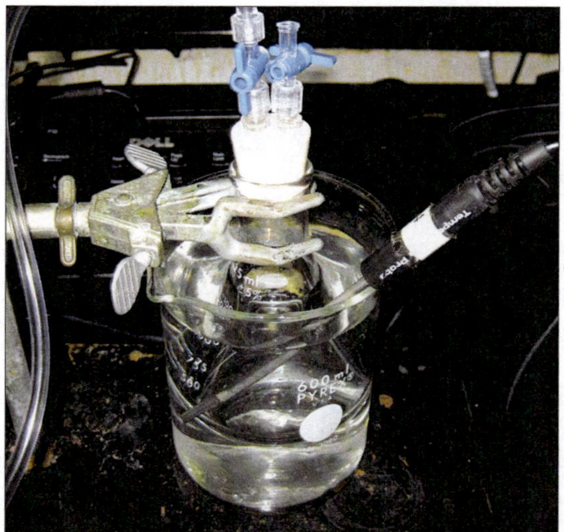

**FIGURE 7.4 Flask with Connectors Clamped into Flask** Temperature probe is submerged in the beaker; connector goes to the Vernier pressure probe.

**11.** There will be a piece of clear tubing with connectors on both ends. Connect one end of the tube to the pressure probe, and connect the other end to one of the stopcocks. ***Be sure both of the stopcocks on the connectors remain closed at this time.***

**12.** Put the temperature probe into the 600-mL beaker.

**13.** Attach the Vernier unit to the computer using the USB cable.

**14.** Log on to the computer. Open the Logger Pro program.

**15.** From the top menu bar, click "file—open—Advanced Chemistry w Vernier—open." Highlight "30b Gases.cmbl," then click "open." The data collection screen for this experiment will open.

**16.** In the lower left corner you will see a box for the pressure measurement and a box for the temperature. Above that are the data points, and a graphical display is to the right.

**17.** The pressure unit will probably be in kPa. Change this to atmospheres. To change the pressure units, you can click on "experiment" from the top menu bar. Select "change units—gas pressure sensor." Select the unit "atm."

**18.** Change the temperature units to degrees Celsius from the same menu.

**19.** Get the water bath up to a temperature of about 37 °C by adding a mixture of tap water and warm water to the beaker. (Hot water is available from the sink near the water distillation apparatus.) If you need to remove water from the beaker at any time, you can do this with a plastic pipette or empty water squirt bottle.

**20.** Once you have everything set up, open the blue stopcock to the pressure probe. (***Do not open*** the other stopcock.)

**21.** You should see the pressure increase. If this does not happen, you need to begin again from Step 6.

**22.** When the pressure and temperature stabilize, click "collect" from the top menu bar. A point will appear on the plot. Click "keep," and the point will appear in the *P-T* table.

**23.** Now add a ***few*** pieces of ice to begin cooling the water bath. Monitor the temperature. When it has gone down 3–5 °C, let it stabilize and collect another data point.

**24.** Continue this process until you get the temperature down to about 12 °C.

**25.** Click "stop" after you collect the final data point.

**26.** The graph should show temperature on the *x*-axis and pressure on the *y*-axis. Scale the graph properly so that it fills up most of the page. Label both axes properly, and put a title on the graph. Get the equation of the line using Logger Pro. Print a copy of the graph for each person. *You will need to attach this graph to your Data Sheet. (If you have forgotten how to work with graphs in Logger Pro, see the handout in the lab on the use of Logger Pro to make graphs.)*

**27.** Close Logger Pro and log off the computer. Please close the computer.

## Clean-Up and Disposal

- Remove the flask assembly from the clamp. Pour out the water in the beaker. Leave the ring stand, clamp, and beaker on the bench top.

- Remove the pressure probe from the connector to the flask. Remove the syringe from the connector and remove the stopper from the flask.

- Wipe the outside of the 125-mL gas flask.

- Disconnect the probes from the Vernier unit.

- ***Leave all of the probes, Vernier units, flasks, connectors, and syringes in the plastic box on the bench top!***

# DATA SHEET

Name: _______________________________    Grade: _______________________________

Date Experiment Performed: _______________    Days Late: _______________________________

CRN of Lab Section: _______________________    Instructor's Initials: _______________________

| General Grading Items | 20 Points |
|---|---|
| Complete Information Readable, Data Recorded, Calculations Shown | |
| Name and CRN Included on Report | |
| All Safety Rules Were Followed | |
| Waste Was Properly Disposed of and Lab Area Was Cleaned | |
| 20-Question Pre-Lab Assignment | /20 |
| **Total** | **/20** |

# Part I: Graphical Analysis of Gas Data

## Graph of Pressure vs. Temperature Data (35 points)

1. Attach a copy of the graph of pressure vs. temperature that you made in lab using Logger Pro. Be sure your name and a date are on the graph and that everything is correctly labeled and scaled.

2. Describe how the pressure of a gas varies with the temperature, if number of moles and volume are kept constant.

3. Write an equation to express this relationship similar to those shown in the introduction section for Boyle's law and Charles' law.

4. From your graph, give the value of absolute zero in units of degrees C. (**Hint:** The pressure at absolute zero would be 0 atm.) *Show your work. Describe how you used the graph to get your answer.*

5. From your graph, give the value of the ideal gas constant in units of L·atm/mol·K.

**6.** The accepted value of the ideal gas law constant is 0.0821 L·atm/mol·K. Calculate the error and % error of the estimate from your graph.

## Graph of Pressure vs. Volume Data (20 points)

**7.** Attach a copy of the graph of pressure vs. volume that you made in lab using Logger Pro. Be sure your name and a date are on the graph and that everything is correctly labeled and scaled.

**8.** Describe how the volume of a gas varies with the pressure, if number of moles and temperature are kept constant.

**9.** What type of fit did you find in Logger Pro that was the best match for the *P-V* data?

**10.** From Logger Pro, what is the equation of the best fit you selected?

**11.** What is the numerical value of the constant in the equation?

**12.** Use the equation from Logger Pro and predict the volume of a gas at a pressure of 3.55 atm.

# PART II: Experimental Determination of *P-T* Relationship and Values for Gas Law Constant and Absolute Zero

## Experimental Determination of Pressure-Temperature Relationship for a Gas (30 points)

**13.** What gas was in the flask?

What are the two main components of this gas? What is the percentage of each component present?

**14.** Calculate the mass of the gas in the flask.

Mass of stopper, flask, and air: _______________ g

Mass of empty flask and stopper: _______________ g

Mass of air in flask: _______________ g

**15.** Assume that the air sample is 78% nitrogen and 21% oxygen. Calculate the average atomic mass of the air sample. (**Note:** You will need to calculate a weighted average.)

**16.** Use the average atomic mass and the actual mass of air in the flask and calculate the number of moles of air in the flask. Show the numbers you used to get your answer.

**17.** Assume that the volume of the flask and tubing system was 135 mL. Convert this value to liters.

**18.** Attach a copy of the graph of pressure vs. temperature that you made in lab using Logger Pro. Make sure your name and a date are on the graph and that everything is correctly labeled and scaled. Be sure the equation for the linear fit is on the graph.

**19. From your graph,** give the value of the ideal gas constant ($R$) in units of L·atm/mol·K. The ideal gas law is $PV = nRT$.

**20. From your graph,** give the value of absolute zero in units of degrees C.

**21.** The actual value of the ideal gas constant is 0.0821 L·atm/mol·K. Calculate the % error of your measured value of $R$.

**22.** The actual value of absolute zero is –273.15°C. Calculate the % error of your measured value of absolute zero.

# Gas Stoichiometry



## Objective

At the completion of the lab, you should be able to

- use the ideal gas law and the principles of stoichiometry to calculate the percent of $NaHCO_3$ in a mixture—a solution of baking soda of unknown concentration. Data for this calculation will be collected by determining the amount of $CO_2$ gas produced in the reaction of HCl with $NaHCO_3$.

# 20-QUESTION PRE-LAB ASSIGNMENT

CRN:_________________________________________     Your Name: _________________________________

Date Submitted: _______________________________     TA's Name: _________________________________

*Please write your answers legibly in the space provided. Many answers will actually consist of three or four pieces of information that are clearly written on the board during the conversation. You must include all aspects of the answer to receive full credit.* **You only need to watch this video from the 19:48 time stamp through to the end to answer this week's pre-lab assignment.**

**1.** Define "gas stoichiometry."

**2.** Write the balanced chemical equation for this week's reaction.

**3.** What type of reaction is this?

**4.** Where does the gas in this reaction come from?

**5.** What is the stoichiometric ratio of $NaHCO_3:CO_2$?

**6.** Draw out and label the experimental apparatus set-up for this week.

**7.** How do we know what the volume of gas produced in this reaction is?

**8.** Write the ideal gas law formula, and then re-arrange it to solve for $n$, the number of moles.

**9.** Which measurements must be converted in order to use them in the ideal gas law?

**10.** Does $P_{atm} = P_{CO_2}$?

**11.** How do we obtain $P_{CO_2}$?

**12.** Which measured pressure is our total pressure, $P_{total}$, in lab this week?

**13.** Which **two** pressures add up to equal $P_{total}$ in our experiment?

**14.** Show the calculation of $P_{CO_2}$ from the video, both in mmHg and converted to atm.

**15.** What does the number of moles of $CO_2$ that were produced in the reaction tell us about the number of moles of $NaHCO_3$ that reacted?

**16.** Write the math equation for the **mass %** calculation?

**17.** Calculate the molar mass of $NaHCO_3$ and report it here to four significant figures. (Only two significant figures were used in the video.)

**18.** The mass % calculated and reported in the video is acceptable and may be close to the value you actually obtain with your data. What was the mass % reported in the video?

**19.** Is the $NaHCO_3$ you use in lab this week toxic?

**20.** What exactly *is* $NaHCO_3$? Identify the manufacturer and the name of the product that they sell it as.

# BACKGROUND

The balanced chemical equation for the reaction involved in this experiment is

$$NaHCO_3(aq) + HCl(aq) \rightarrow NaCl(aq) + H_2O + CO_2(g)$$

In this reaction, for every one mole of sodium hydrogen carbonate ($NaHCO_3$) that reacts, one mole of carbon dioxide gas ($CO_2$) is produced. Thus, the amount of $CO_2$ gas produced in the reaction can be used to determine the amount of $NaHCO_3$ that is present in the original sample. (**N.B.,** Other names for $NaHCO_3$ are sodium bicarbonate or baking soda.)

In this experiment, it is more convenient to measure the volume ($V$) of $CO_2$ gas that is produced in the reaction. Then we can relate the volume of gas that is produced to the number of moles of gas produced using Avogadro's law for gases. One of the consequences of Avogadro's law for gases is that the volume of a gas present at constant temperature ($T$) and pressure ($P$) is directly proportional to the number of moles of gas present. If "$n$" represents the number of moles of a gas present, the Avogadro's law for this example is

**$n$ is directly proportional to $V$ (at constant $T$ and P),   or   $n \propto V$**

Recall other gas laws:

- Charles' law: $V \propto T$
- Boyle's law: $V \propto 1/P$
- Gay-Lussac's law: $P \propto T$

All four gas laws can be combined to produce the ideal gas law

**$PV = nRT$**

The symbol "$R$" is called the ideal gas constant with a value of 0.08206 L·atm/mol·K. By using this value for "$R$," the pressure must be expressed in units of atmospheres (atm); the volume must be expressed in units of liters (L); and the temperature must be expressed in kelvin units (K). The relationship in the ideal gas law summarizes the behavior of an ideal or perfect gas. No real gas is ideal, but real gases behave almost like ideal gases under appropriate conditions (like this experiment) to treat them as ideal.

The ideal gas law can be rearranged as follows to solve for moles of gas:

**$n = PV/RT$**

The variables $T$, $V$, and $P$ will be measured in the lab.

## Temperature

Since the experiment is performed at room temperature, the temperature of the gas that is produced will simply be the room temperature of the laboratory today.

# Volume

To determine the volume of gas produced, first consider the experimental set-up to capture the gas. The gas produced in the reaction will occupy a certain volume. To accommodate this extra volume, the gas that is produced from the reaction in the first flask will displace an equal volume of salt water solution from the second flask. The salt water solution that is displaced can be collected in a beaker, and its volume can be measured with a graduated cylinder. The volume of salt water solution that is displaced is equal to the volume of gas produced.

# Pressure

To determine the pressure of the $CO_2$ gas that is produced, we must consider Dalton's law for gases. Dalton's law of partial pressures states that the total pressure exerted by a mixture of gases is equal to the sum of the partial pressures exerted by each of the separate gases.

$$P_{total} = P_1 + P_2 + P_3 + \dots$$

In our system the carbon dioxide gas is collected by bubbling it through water. So, the gas that is collected contains not only carbon dioxide gas, but water gas (water vapor) as well. The partial pressure of water vapor over a saturated sodium chloride (salt) solution at a given temperature is known and can be found in an appropriate source. Since the gas is collected under the conditions in the laboratory, the total pressure of the gas collected is simply the atmospheric pressure. Dalton's law can be applied to this situation to determine the pressure of the carbon dioxide gas that is produced.

$$P_{total} \equiv P_{atm} = P_{CO_2} + P_{H_2O}$$

$$P_{CO_2} = P_{atm} - P_{H_2O}$$

**Sample problem:** The temperature and pressure in lab were determined to be 22°C and 765 mmHg, respectively. An experiment was performed to produce carbon dioxide gas. The gas was collected by bubbling it through salt water solution and displacing an equivalent volume of salt water from an inverted test tube. The amount of salt water displaced by the gas was exactly 200.0 mL. It is known that the partial pressure of water over a saturated NaCl (salt) water solution at 22°C is 15 mmHg. Calculate the moles of carbon dioxide gas, $CO_2$, that were produced in this experiment.

**Solution:**

$$T = 22°C + 273 = 295 \text{ K}$$

$$V = \frac{200.0 \text{ mL}}{} \cdot \frac{1 \text{ L}}{1000 \text{ mL}} = 0.200 \text{ L}$$

$$P_{CO_2} = P_{total} - P_{H_2O} = 765 \text{ mmHg} - 15 \text{ mmHg} = 750 \text{ mm Hg}$$

$$P_{CO_2} = \frac{750 \text{ mm Hg}}{} \cdot \frac{1 \text{ atm}}{760 \text{ mm Hg}} = 0.987 \text{ atm}$$

$$n = \frac{PV}{RT} = \frac{(0.987 \text{ atm})(0.2000 \text{ L})}{(0.08206 \text{ L·atm/mol·K})(295 \text{ K})} = 8.16 \times 10^{-3} \text{ mol } CO_2 \text{ gas}$$

# Percent Composition by Mass

Recall that a percentage is the part of something present divided by the whole, and then multiplied by 100%.

**percent (%) = (part/whole) × 100%**

Percent composition by mass indicates the percentage of the total mass of a sample that is due to one particular component of the sample. In this case, the "part" of the sample is the mass of the particular component, and the "whole" is the mass of the entire sample.

$$\% \text{ by mass} = \frac{\textbf{mass of component}}{\textbf{mass of sample}} \times \textbf{100\%}$$

**Sample Problem:** 126.3 g of solution was found to contain 17.8 g of calcium carbonate, $CaCO_3$. Calculate the percent by mass of $CaCO_3$ in the sample.

**Solution:**

$$\% \ CaCO_3 = \frac{17.8 \text{ g } CaCO_3}{126.3 \text{ g solution}} \times 100\% = 14.1\% \ CaCO_3$$

# Materials

- graduated cylinder
- 500-mL Erlenmeyer flask
- Florence flask
- 2 small test tubes
- 2-hole stoppers
- rubber bulb
- clamp
- beaker

The experimental set-up that will be used can be seen in Figure 8.1. Step-by-step directions for the experimental set-up with be given in the Procedure.

**FIGURE 8.1**

**TABLE 8.1  Vapor Pressure of Water and Salt Solutions**

| $T$ (°C) | $P_{H_2O}$ in mmHg | |
| --- | --- | --- |
| | Pure $H_2O$ | NaCl Solution |
| 15 | 12.8 | 9.7 |
| 16 | 13.6 | 10.3 |
| 17 | 14.5 | 11.0 |
| 18 | 15.5 | 11.7 |
| 19 | 16.5 | 12.4 |
| 20 | 17.5 | 13.2 |
| 21 | 18.7 | 14.1 |
| 22 | 19.8 | 15.0 |
| 23 | 21.1 | 15.9 |
| 24 | 22.4 | 16.9 |
| 25 | 23.8 | 17.9 |
| 26 | 25.2 | 18.9 |
| 27 | 26.7 | 20.0 |
| 28 | 28.3 | 21.0 |

# PROCEDURE

1. Place a 500-mL Erlenmeyer flask on the balance and tare the balance to 0.00 g.

2. Add unknown solution to the weighed 500-mL Erlenmeyer flask until about 10 g of the solution have been added. Approximately accurately record the mass of the solution in Table 8.2.

3. Add about 40 mL of water to the Erlenmeyer flask.

4. Fill the small test tube within ½ inch of the top with 4 M HCl solution. Carefully slide the test tube containing HCl solution into Flask A. Avoid spilling any HCl into the unknown solution. (If HCl is spilled into the unknown solution, the experiment must be started over.)

5. Place salt water in the Florence flask until it is just over half full. Place the two-hole stopper with tubing into the flask. Place the end of Tube 2 into the beaker.

6. Using a rubber bulb, blow air through Tube 1 to force salt solution into Tube 2. If necessary, use suction from the bulb to pull salt water back into Tube 2. Once you are confident that Tube 2 is filled with water, place a clamp on the tube to keep the salt water in place.

7. Place a stopper firmly into the Erlenmeyer flask and connect Tube 1. Be careful not to overturn the small test tube in the flask at this time.

8. Remove the clamp from Tube 2 (the end of Tube 2 is resting in salt water in the beaker), and raise the beaker so that the salt solution levels in the beaker and Florence flask are equal. (The pressure in the two flasks then equals atmospheric pressure.) Replace clamp on Tube 2, and empty the excess salt solution from the beaker into the appropriate storage container. (Do not discard the salt water solution; it is being recycled and reused.)

9. At the same time, carefully tilt the Erlenmeyer flask to overturn the test tube of HCl and **undo the clamp on Tube 2.** The HCl solution will react with the $NaHCO_3$ in your sample. Gently shake the Erlenmeyer flask, holding it by the neck until all the $CO_2$ has been evolved.

10. Salt solution should partially fill the beaker (remember this represents the volume of gas that is produced). When all the $CO_2$ has been evolved, again raise the beaker so that the salt solution levels in the beaker and Florence flask are equal (the total pressure of gases in the flasks is then equal to atmospheric pressure) and secure the clamp on Tube 2.

11. Using a graduated cylinder, measure the volume of salt solution that was collected in the beaker. Record the volume in Table 8.2.

12. Using your data, calculate the percent by mass of $NaHCO_3$ in the sample. **N.B.,** The formula weight for $NaHCO_3$ is 84.01 g/mol.

13. Using a fresh 10-gram aliquot of the same unknown $NaHCO_3$ salt solution, repeat the procedure to obtain another set of data. You must use a fresh sample of HCl; appropriately discard the solution that was produced in the Erlenmeyer flask.

| **TABLE 8.2  Experimental Data** | | |
| --- | --- | --- |
| | **Trial 1** | **Trial 2** |
| Temperature (°C) | | |
| Total Pressure of Gas Collected; Atmospheric Pressure (mmHg) | | |
| Mass of Sample Solution | | |
| Volume of Salt Water Solution Displaced | | |

# DATA SHEET

Name: _______________________________________     Grade: _______________________________________

Date Experiment Performed: ___________________     Days Late: ___________________________________

CRN of Lab Section: _________________________     Instructor's Initials: _______________________

| General Grading Items | 20 Points |
|---|---|
| Copies of Lab Pages Submitted; Labeled with Name and Date, Complete Information, Readable, Data Recorded Matches Results Given in Report | |
| Name and CRN Included on Report | |
| All Safety Rules Were Followed | |
| Waste Was Properly Disposed of and Lab Area Was Cleaned | |
| Evaluation of Student Performance Overall (Student Was on Time, Followed Safety Rules, Performed the Lab Correctly and Within the Time Allowed, Etc.) | |
| 20-Question Pre-Lab Assignment | /20 |
| **Total** | **/20** |

**CHE 111L student: Please submit a copy of your lab notebook pages with today's data to your TA before you leave.**

You must show at least one set of calculation set-ups (there is room available on page 171).

**TABLE 8.3**

| | Trial 1 | Trial 2 |
|---|---|---|
| Temperature (°C) | | |
| Temperature (K) | | |
| Atmospheric Pressure of Gas Collected (mmHg) | | |
| $P$ of Water Vapor over Saturated NaCl Solution (mmHg)—From Table | | |
| Pressure of $CO_2$ (mmHg) | | |
| Pressure $CO_2$ (atm) | | |
| Mass of Sample (g) | | |
| Volume of $CO_2$ Gas Produced (mL) | | |
| Volume of $CO_2$ Gas Produced (L) | | |
| Moles of $CO_2$ Produced | | |
| Moles of $NaHCO_3$ Reacted | | |
| Grams of $NaHCO_3$ Reacted | | |
| % $NaHCO_3$ in Sample | | |

Please show one full set of calculations below:

# Calorimetry
## The Determination of the Specific Heat of a Metal

## Objective

At the completion of the lab, you should be able to

- determine the specific heat of a metal by performing a **constant pressure calorimetry** experiment.

## Reference

Silberberg and Amateis, 8e, pages 258–261, 268–271.

# 20-QUESTION PRE-LAB ASSIGNMENT

CRN:_______________________________________     Your Name: ___________________________________

Date Submitted: _________________________________     TA's Name: ________________________________

*Please write your answers legibly in the space provided. Many answers will actually consist of several pieces of information that are clearly written on the board during the conversation. You must include all aspects of the answer to receive full credit.*

**1.** How do we maintain constant pressure in the laboratory when we are conducting an experiment?

**2.** What does the science of calorimetry study?

**3.** In what direction does heat *always* flow?

**4.** Sketch and label a coffee cup calorimeter.

**5.** Sketch and label the boiling water bath.

**6.** What is the warmer body this week? What is the cooler body?

**7.** How do we bring these two bodies into direct contact with each other?

**8.** What do we name the same temperature that both objects eventually achieve?

**9.** What is important about the above temperature?

**10.** Write down the calorimetry equation and identify each variable.

**11.** In the sample calculation for Trial 1, how much heat was transferred into the water? Where did that heat come from?

**12.** What is the specific wording for the law of energy conservation for a heat transfer process?

**13.** In our sample calculations, how are the numerical values and the signs (+ or –) of the amounts of heat transferred ($q_w$ and $q_m$) related to each other? Show these relationships using $q_w$ and $q_m$ from the calculation for Trial 2.

**14.** What are the only types of numerical values that should be averaged together?

**15.** When we talk about precision, what exactly are we comparing to each other?

**16.** Why is it called the **±5% rule,** rather than, say, the **±6% rule?**

**17.** Write down the law of Dulong and Petit. About when did they "discover" it? How did they deduce it?

**18.** How can we write the number 200 to two significant figures? Show it below.

**19.** Why had our sample calculation metal better **not** be thallium?

**20.** What equation do we use to assess accuracy? And what exactly are we comparing when we assess accuracy? Finally, what is the acceptable numerical range within which two values are said to be accurate?

## TABLE 9.1  Sample Calculation Data

| | Trial 1 | Trial 2 |
|---|---|---|
| Mass of Water ($m_{water}$) | 21.40 g | 30.08 g |
| Mass of Metal ($m_{metal}$) | 48.00 g | 78.93 g |
| Boiling Water Temperature ($T_{high}$) | 100.0°C | 98.7°C |
| Initial Temperature of Water in Calorimeter ($T_{low}$) | 19.8°C | 16.3°C |
| Final Temperature of Water and Metal in Calorimeter ($T_{eqbm}$) | 25.0°C | 22.6°C |
| $\Delta T_{water} = T_{eqbm} - T_{low}$ | | |
| $\Delta T_{metal} = T_{eqbm} - T_{high}$ | | |
| Specific Heat of Water ($sh_{water}$) | 4.184 J/g°C | 4.184 J/g°C |
| Heat Transferred to Water ($q_{water}$) | | |
| Heat Transferred from Metal ($q_{metal}$) | | |
| Specific Heat of Metal ($sh_{metal}$) | | |
| Precision of Your $sh_{metal}$'s | | |
| Average Specific Heat of Metal | | |
| Molar Mass of Metal via Dulong and Petit | | |
| % Error of Your Empirical Molar Mass | | |

# BACKGROUND

**Temperature** is described in the kinetic molecular theory of gases as a measure of the average kinetic energy of all the particles in a system, $T \propto KE_{avg}$.

The **system** is the portion of the universe that is of poignant interest to us. The system usually includes **only** the chemicals that a chemist is mixing together; it does **not** typically include the container that surrounds the chemicals.

**Heat** is a form of energy (thermal energy, the energy associated with the random motion of atoms, and molecules) that *always* flows spontaneously from a region of higher temperature to a region of lower temperature (the Zeroth law of thermodynamics).

Two objects that start out at different temperatures will eventually reach the same intermediate temperature (the equilibrium temperature, $T_{eqbm}$) when they are brought into contact with each other.

**Temperature change** is therefore a logical aspect of the system for us to monitor as we study heat flow, the exchange of heat between objects in the system.

**Calorimetry** is the branch of thermodynamics that studies heat (ex)change, and **thermodynamics** is the scientific study of the interconversion of heat and *other* kinds of energy.

Heat flow may be measured in a device called a calorimeter. A **calorimeter** is a vessel which has insulating walls and therefore is **not** a part of the system. Within the boundaries of the calorimeter itself, heat may transfer from one part of the system to another; however, **no** heat flows into or out of the calorimeter from the surroundings. In this week's experiment, a simple calorimeter made from two nested Styrofoam cups will be used. This apparatus is colloquially called a "coffee cup calorimeter," and is illustrated for you in Figure 6.9 on page 269 in Silberberg and Amateis.

Inside the coffee cup calorimeter, two objects will be brought into direct thermal contact with each other, and the heat flow will be confined to be between them only. (Heat will flow from the warmer metal shot to the cooler water, and the metal shot will cool to the same temperature that the water will warm up to ($T_{eqbm}$) as thermal equilibrium is achieved between the two objects.)

## Specific Heat

Specific heat is an **intensive** physical property of matter. (Recall that **in**tensive physical properties are **in**dependent of the amount of matter present.) Like density, specific heat is different for different materials. It is also a proportionality constant that relates heat and temperature change in the following way.

Heat flow into a substance causes the temperature of that substance to rise; and, similarly, heat flow out of a substance causes a decrease in temperature. The amount of heat ($q$) associated with a temperature change ($\Delta T = T_{final} - T_{initial}$) is directly proportional to the mass ($m$) of the substance and to $\Delta T$. This relationship is expressed in the calorimetry equation.

**Calorimetry equation:**

$$q = (m)(sh)(\Delta T)$$

Specific heat is thereby defined as the amount of heat (energy) required to raise the temperature of one gram of a substance by one degree Celsius (which is the same as raising the temperature by one kelvin unit). The units of heat may be calories or Joules, with your textbook authors preferring Joules. The units of specific heat can be cal/g°C or J/g°C. The traditional unit of energy is the calorie, and the SI unit is the Joule. So, 1 calorie of heat energy is exactly equal to 4.184 J, and the conversion factor is 1.000 cal = 4.184 J. The specific heat of water is usually reported to three significant figures as

**1.000 cal/g°C   or   4.184 J/g°C**

The specific heats of several substances are reported to three significant figures in Table 6.2 on page 268 in Silberberg and Amateis. The specific heat of water contains four significant figures.

When two isolated objects at different initial temperatures **and** in direct contact exchange heat, the thermal energy flow is governed by both the Zeroth law of thermodynamics (heat **always** flows from a warmer body to a cooler body) and the first law of thermodynamics (the law of energy conservation). However, rather than stating that "heat is neither created nor destroyed, but only changed in form," the better calorimetry statement is that "all the heat lost by the warmer object is gained by the cooler object." To a first approximation we can represent this heat transfer by the equation

$$q_{warmer} = q_{cooler}$$

and, if we pay heed to the sign convention, the true relationship is better expressed as the pseudo-algebraic equality

$$-q_{warmer} = +q_{cooler}$$

since the warmer body **loses** heat, and the cooler body **gains** heat. The heat change for each object is identical in magnitude, but opposite in sign.

In this experiment, the specific heat of a metal will be calculated by heating a weighed sample of metal in a boiling water bath ($T_{high} = T_{i,metal}$) and then adding the hot metal to a measured mass of cooler, room temperature water already inside the coffee cup calorimeter ($T_{low} = T_{i,water} \approx T_{room\ temp}$). As heat flows out of the metal and into the water, both objects approach the equilibrium temperature ($T_{eqbm} = T_{final}$). Therefore, $+q_{cooler}$ ($q_{water}$) can be calculated for the water; and, since $+ q_{cooler} = -q_{warmer}$, $q_{warmer}$ ($q_{metal}$) can be used to calculate $sh_{metal}$.

A sample set of data will be provided to you in pre-lab lecture, and we will run through a sample calculation for that data set.

# The (empirically determined) Law of Dulong and Petit

Further, for metals, there is an empirical relationship that was gleaned in the early-1800s and which can be expressed as

$$(sh_{metal})(\text{molar mass}) \approx 26 \text{ J/mole}°\text{C}$$

Called the law of Dulong and Petit, it can be used to determine the molar mass of an unknown metal by dividing 26 by the empirically determined specific heat ($sh_{metal}$). (You should do the dimensional analysis to convince yourself that you will end up with the correct units of g/mole.)

The result of this calculation can be compared to the atomic weights for metals on the periodic chart such that it may be possible for you to determine which metal you used in this experiment. (Select the metal whose atomic weight on the periodic chart is closest to the empirical molar mass you obtain by using the law of Dulong and Petit.)

Alternatively, once you know the identity of your metal, you can calculate a percent error for the work you did in lab this week by comparing your empirical molar mass value to the true molar mass value.

# PROCEDURE

***Safety glasses must be worn while working in the laboratory!***

1. Set up a boiling water bath by placing a 400-mL beaker ⅔ full with water on the ring clamp with wire gauze (or a clay triangle) under it. While doing the next two steps, heat the water to boiling.

2. Place the calorimeter (two nested Styrofoam cups) on the balance and tare the balance. Add approximately 30–40 grams of water into the calorimeter. Accurately record the mass of water on the Data Sheet.

3. Weigh out approximately accurately 35 g of metal shot. Record this mass accurately on the Data Sheet. Carefully and slowly pour the metal shot into a test tube. Tighten a ring stand clamp near the top of the test tube. This will be used as a "handle" when you work with the hot test tube.

4. Gently submerge the test tube containing the metal shot into the boiling water bath. The surface of the metal shot should sit approximately one inch below the surface of the water. Allow the metal-shot-containing test tube to sit in boiling water for at least five minutes. At the end of five minutes, measure and record the temperature of the metal shot inside the test tube to the tenths place. Perform Step 5 while the test tube remains in the boiling water bath.

5. Remove the temperature probe from the metal shot, cool it under running tap water, then dry it off. Let it sit on the tabletop for a minute or two. Insert the temperature probe through the calorimeter lid, and immerse the temperature probe into the water in the calorimeter. Carefully swirl the water around the temperature probe (being careful not to spill any). Measure and record this initial water temperature to the tenths place on the Data Sheet.

6. Using the ring stand clamp, remove the test tube from the boiling water, and blot off any excess water on the outside of the tube. Quickly, but carefully (no splashing), pour the metal all at once into the calorimeter. Immediately replace the lid and swirl ever so gently. Observe the rise in temperature, and record the maximum temperature obtained in the calorimeter as the final (equilibrium) temperature on the Data Sheet.

7. Decant the water out of the calorimeter, pour your metal back into the test tube, and place the test tube back into the boiling water bath. Repeat Steps 2–6 for Trial 2.

## Measurements and Calculations

For each trial, you will measure and report five data—$m_{water}$, $m_{metal}$, $T_{high}$, $T_{low}$, and $T_{eqbm}$. Then you will calculate $\Delta T_{water}$, $\Delta T_{metal}$, $q_{water}$, $q_{metal}$, and $sh_{metal}$. ***Pay close attention to the sign convention!***

Then you will calculate an average $sh_{metal}$, from which you will calculate a molar mass using the law of Dulong and Petit. Finally, you will calculate the precision of your $sh_{metal}$'s, and the % error of your molar mass vs. the true value molar mass of your metal. Your TA will tell you the identity of the metal after you have collected your data.

# DATA SHEET

Name: ___________________________________     Grade: ___________________________________

Date Experiment Performed: ___________________     Days Late: ___________________________

CRN of Lab Section: _______________________     Instructor's Initials: _______________________

**Metal recovery:** The metal shot is fairly expensive and must be recovered for re-use. Please pour your metal shot into the bins provided. ***Do not mix*** one type of metal shot with the other!

Return the Vernier unit, along with the probe and all cables, to the plastic box on the bench top.

| TABLE 9.2 Data Table | Trial 1 | Trial 2 |
|---|---|---|
| Mass of Water ($m_{water}$) | | |
| Mass of Metal ($m_{metal}$) | | |
| Boiling Water Temperature ($T_{high}$) | | |
| Initial Temperature of Water in Calorimeter ($T_{low}$) | | |
| Final Temperature of Water and Metal in Calorimeter ($T_{eqbm}$) | | |
| $\Delta T_{water} = T_{eqbm} - T_{low}$ | | |
| $\Delta T_{metal} = T_{eqbm} - T_{high}$ | | |
| Specific Heat of Water ($sh_{water}$) | 4.184 J/g°C | 4.184 J/g°C |
| Heat Transferred to Water ($q_{water}$) | | |
| Heat Transferred from Metal ($q_{metal}$) | | |
| Specific Heat of Metal ($sh_{metal}$) | | |
| Precision of Your $sh_{metal}$'s | | |
| Average Specific Heat of Metal | | |
| Molar Mass of Metal via Dulong and Petit | | |
| % Error of Your Empirical Molar Mass | | |

# Atomic Structure, Periodicity, Waves, and Emission Spectroscopy

# 20-QUESTION PRE-LAB ASSIGNMENT

CRN:_________________________________________   Your Name: _______________________________

Date Submitted: _______________________________   TA's Name: _______________________________

*Please write your answers legibly in the space provided. Many answers will actually consist of several pieces of information that are clearly written on the board during the conversation. You must include all aspects of the answer to receive full credit. **(You only need to watch this video through the 30:20 time stamp to answer this week's pre-lab assignment.)***

**1.** What is the overlying theme of this week's lab, and what are we paying attention to about that theme?

**2.** What is the Bohr-Rutherford model also known as? Why?

**3.** What is the main point you should take from the Bohr-Rutherford model?

**4.** What is $V$ this week? Write the $V$ equation below and identify $k$, $Q_1$, $Q_2$, and $d$.

**5.** What are the proportionalities from the $V$ equation?

**6.** Write out the law of periodicity. Who deduced it? When?

**7.** List the three periodic trends we investigate this week.

**8.** Write out the full analogy used in the video to get you thinking about core charge.

**9.** Distinguish **core electrons** vs. **valence electrons.**

**10.** Why are the noble gases "noble"? What are their **magic numbers** of electrons?

**11.** Write down the general formula to calculate a core charge, and then use it to calculate a core charge for calcium.

**12.** What is the basis for presenting atomic size as atomic radius?

What is $r$'s (the radius') relationship to core charge?

**13.** Define our third trend. Is this process endothermic or exothermic?

**14.** Write the thermochemical equation for ionization energy, and identify all the entities in the equation.

**15.** What is the relationship between the trends of atomic radius and ionization energy?

**16.** Draw and label the wave that starts at the origin. What do we call this wave?

Identify and define $\lambda$ and $\nu$ for any wave.

What is the frequency of a Gulf Coast water wave? What is a Hertz?

**17.** Show how we get the units for $c$, the speed of light.

**18.** Write the proportionality and then the equivalency that exists between energy and frequency. What is **h?**

**19.** Draw and label the sketch of an "atom"—its nucleus and first three energy levels/shells—
to show how emission spectroscopy can be explained.

Does an emission spectrum look like a continuous rainbow?

What **does** the emission spectrum appear as?

**20.** Why is the element helium called "helium"?

# BACKGROUND

This lab covers information from Chapters 7 and 8 in your textbook. The concepts in this chapter are sometimes difficult for nascent scientists to visualize. By having you do some "virtual" lab experiments, look at some animations, and perform several calculations, we hope that you will be better able to visualize and understand the abstract concepts in this chapter. There are five sections to this lab:

1. Attractive Potentials

2. Basic Concepts of Atomic Structure, Electron Configuration, and Periodicity

3. The Electromagnetic Spectrum

4. Electromagnetic Radiation and Atoms

5. Quantum Numbers, Orbital Shapes, and Electron Configurations

You should fill in the Data Sheet during your lab period by working with your lab partner, other pairs around your work station, and with your TA when necessary. You will need to visit the websites listed and use your text to complete some sections of the virtual lab experience. You may also consult other chemistry or physics texts, other websites, or any other resources to obtain the needed information.

# PROCEDURE

## Part I: Attractive Potentials

Consider two oppositely charged particles with charges $q_1$ and $q_2$ separated by a distance, $d$.

Opposite charges attract, and the energy of this attractive interaction is described by the Coulombic Potential, $V$,

$$V = k \; \frac{q_1 \, q_2}{d}$$

where $k$ is a constant.

This equation tells you that as the distance between two charged particles increases, the energy decreases; and as the charges increase, the energy increases.

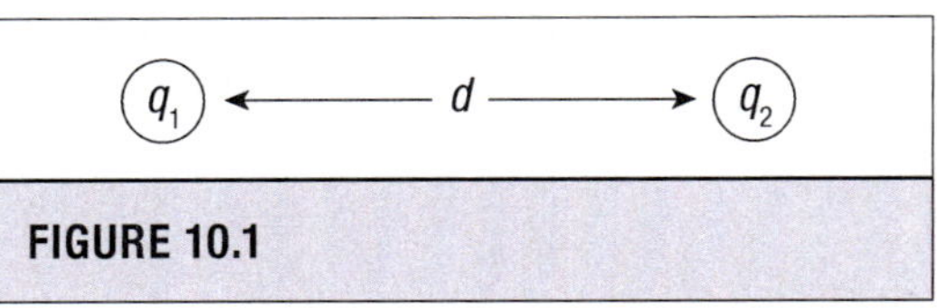



FIGURE 10.1

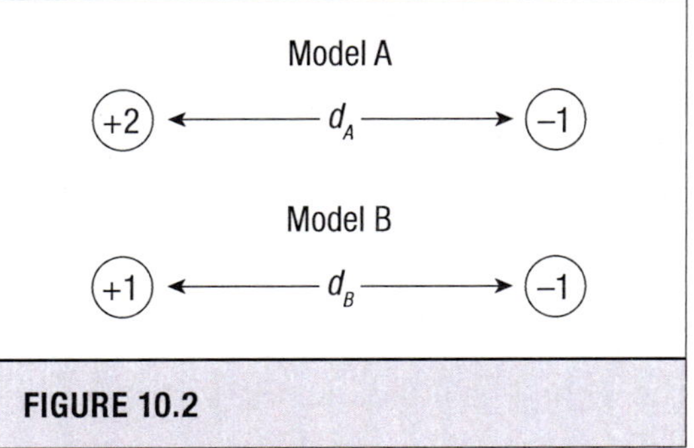



FIGURE 10.2

Consider the two models in Figure 10.2, then answer Questions 1 and 2 on the Data Sheet.

This same model gives insight to the structure of atoms. The Coulombic Potential describes the attraction of negatively charged electrons to a positively charged nucleus and can be used to predict trends such as core charge, atomic radii, ionization energies, and other periodic trends.

## Part II: Basic Concepts of Atomic Structure, Electron Configuration, and Periodicity

Atoms are composed of a nucleus containing protons and neutrons with electrons surrounding the nucleus. In an electrically neutral atom, the number of protons in the nucleus must be exactly equal to the number of electrons that surround it. The most basic model of the atom is Bohr's "solar system" or shell model. Bohr envisioned the hydrogen atom as a nucleus (analogous to our sun) of charge +1 with a single electron (analogous to a planet) orbiting the nucleus at some fixed distance. These orbits are well defined circles, and only certain orbits (and, hence, certain radii) are allowed. In this model, as we progress from hydrogen to helium and beyond, we view the additional electrons as being added to successively larger "shells."

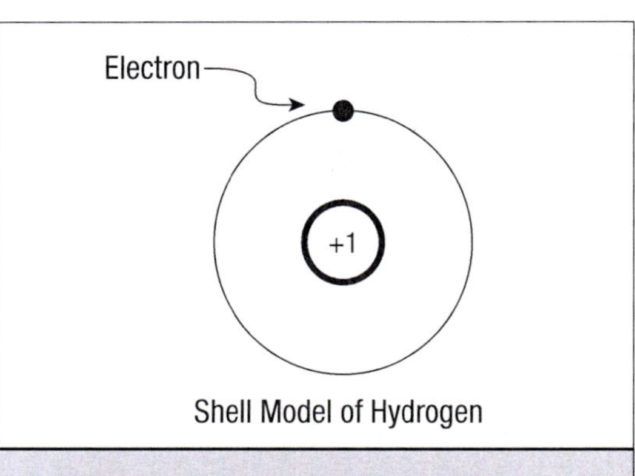



FIGURE 10.3 The Electron "Looks" to the Inside of the Atom and "Sees" a Charge of +1

Consider the shell model of helium, then answer Questions 3 and 4 on the Data Sheet.

To look at some animated versions of the shell model, go to *http://video.sunflowerlearning.com/browse/productvideos/video/atomsandions*.

The video runs on its own because this is just a product preview, but you can watch the electrons that swirl around the nuclei of hydrogen, helium, lithium, and calcium. You can pause the video at any time and fast forward, etc., as needed.

There is an alternative website (David's Whizzy Periodic Table) that you can visit if your own PC

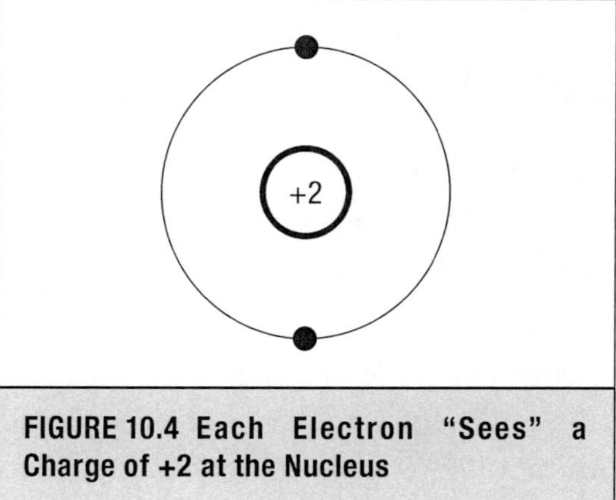

**FIGURE 10.4 Each Electron "Sees" a Charge of +2 at the Nucleus**

or MAC has the updated version of Adobe. This website is not accessible from EKU computers but is described in the appendix at the very end of this lab handout for those of you who would like to watch these animations on your own PC or MAC.

Answer Questions 5 and 6 on the Data Sheet. You will not be able to complete the last three columns in Table 10.2 at this point.

## Core Charge

Consider the shell model of lithium. Notice that in this model the electron that is easiest to remove is further from the nucleus than the two inner electrons in the first shell. All electrons have the same negative charge and repel each other. The repulsion of the outer electrons by the inner electrons dramatically reduces the attraction of the outermost electron to the nucleus. Thus the **net** charge acting on the outermost electron is the net charge of the nucleus and the inner shell or core electrons. For lithium this net charge is +1, resulting from +3 nuclear charge and a –2 core electron charge, i.e., +3 + (–2) = +1. In general the nucleus plus the inner shell electrons is referred to as the **core** of an atom, and its net charge is the core charge. The outermost shell of an atom is referred to as the **valence** shell, and the electrons contained in this shell are the valence electrons.

Now go back and complete the remaining three columns in Table 10.2, and answer Questions 7–14 on the Data Sheet.

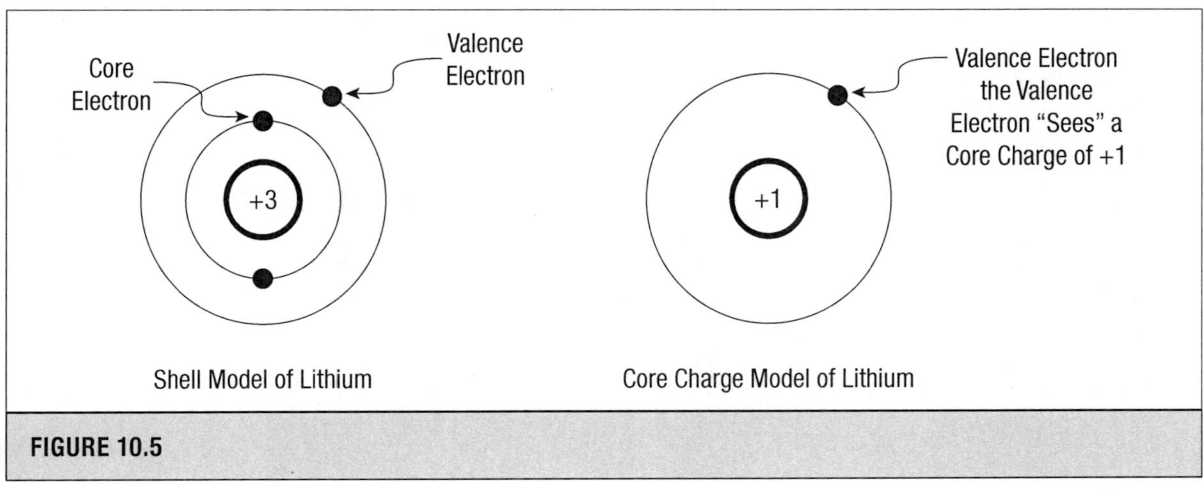


**FIGURE 10.5**

## Part III: The Electromagnetic Spectrum

Electromagnetic radiation is a type of energy that originates from electron transitions in atoms. *Go to www.youtube.com/watch?v=kIdESF0OwTo* and watch "Physics-Waves in the Real World: Properties of 3D waves" from the 45-second mark through the 1-minute, 38-second mark (0:45–1:38) to watch an example of a wave.

A more elaborate animation to help you understand the concept of electromagnetic radiation is at *http://www.youtube.com/watch?v=aCTRjVEmeC0* entitled "Electromagnetic Waves Animation." It shows you that electromagnetic radiation consists of electric and magnetic waves vibrating at right angles to each other. Answer Questions 15–19 on the Data Sheet.

### Regions of the Electromagnetic Spectrum

The characteristics of electromagnetic radiation depend on its energy. To see the different regions of the electromagnetic spectrum, go to *http://teachers.stupidchicken.com/files/active/0/EMFlash.swf.* Answer Questions 20 and 21 on the Data Sheet.

## Part IV: Electromagnetic Radiation and Atoms

### The Hydrogen Spectrum

Electromagnetic radiation and atomic structure are related to each other. UV, visible, and IR electromagnetic radiation originate from the transitions of electrons in atoms. The easiest atom to begin with is hydrogen because it has only one electron. To see electronic transitions give rise to lines in the emission spectrum go to *www.youtube.com/watch?v=Bedn4h_B8WM.*

In Part I you looked at diagrams of atoms using concentric circles to show the electron energy levels. A more common way to show the electron energy levels is by using horizontal lines. The lowest energy level is at the bottom, and the highest is at the top. We can apply energy to a sample of atoms using electricity. This results in electrons being excited from their ground state to higher energy levels. When the electrons de-excite by "falling" to lower energy levels, the energy they lose is emitted in the form of electromagnetic radiation.

To see both types of diagrams for the hydrogen atom, go to *http://www.daviddarling.info/encyclopedia/H/hydrogen_spectrum.html.*

Use these diagrams to answer Questions 22–24 on your Data Sheet.

The website below provides us with another opportunity for us to view the shells in Bohr's model, the hydrogen emission spectrum, and the Lyman, Balmer, and Paschen series of electronic transitions.

*http://chemwiki.ucdavis.edu/Textbook_Maps/General_Chemistry_Textbook_Maps/Map%3A_Chem1_(Lower)/05._Atoms_and_the_Periodic_Table/The_Bohr_Atom*

The Balmer series falls within the visible region of the electromagnetic spectrum. If you used a diffraction grating or prism to break up the light that comes from an excited hydrogen sample into its colors, you would see the four lines that represent the hydrogen spectrum.

*http://chemed.chem.purdue.edu/genchem/topicreview/bp/ch6/bohr.html*

Choose the "default" button to see a drawing of the apparatus used to obtain the emission spectrum of hydrogen and the spectrum itself complete with wavelengths reported for the lines in the spectrum.

Click on "H," and you will see the emission spectrum for hydrogen in the area above the periodic table. If you click and hold on the spectral lines, the wavelength of the lines will appear at the bottom of the line.

(**Note:** These wavelengths are in units called Angstroms. To convert to nm, divide by 10. For example, a wavelength of 5255 on this chart would be equal to 525.5 nm.)

Below, make a list of the hydrogen lines you see. Give the color and wavelength, in nm.

**TABLE 10.1**

| Color of Line | Wavelength, in nm |
|---|---|
|  |  |
|  |  |
|  |  |
|  |  |

## Spectra from Other Elements

Any element can be put into a glass tube and electrically excited. The atoms from the element will emit a spectrum that is unique for each element. This is one way you can identify what elements are in a sample of material.

To see some emission spectra, go to *http://chemistry.bd.psu.edu/jircitano/periodic4.html.*

You will see a periodic table. Click on any element to see its emission spectrum, then answer Questions 25 and 26 on the Data Sheet.

You are welcome to stop here because you have completed the virtual lab activity that is required.

However, if you would like to continue to learn about the quantum mechanical model of the atom, the page that follows will assist you in learning about that theory.

# Part V: Quantum Numbers, Orbital Shapes, and Electron Configurations

The modern version of atomic theory is more complex than the simple Bohr atom, which showed electrons revolving around the nucleus the way planets revolve around the sun. Now we use the *quantum mechanical model* of the atom. A physicist named Erwin Schrödinger developed equations that would describe the *probability* of finding an electron in an area of space around the nucleus. These equations have four important numbers in them, referred to as quantum numbers. They are:

1. The principal quantum number ($n$): This gives information about the size (distance away from the nucleus) and energy level of an orbital.

2. The angular momentum quantum number ($l$): This gives information about the three-dimensional shape of an electron orbital.

3. The magnetic quantum number ($m$): This gives information about the location of an electron orbital along (or between) the $x$-, $y$-, or $z$-axes in 3D space.

4. The spin quantum number ($s$): This gives information about the rotation of the electron about its own axis—spin "up" or spin "down."

   When the quantum mechanical equations are solved, we can get information about the probable location of electrons in space around a nucleus.

Answer Question 27 on the Data Sheet.

## The Angular Momentum Quantum Number

The possible values of $l$ range from 0 to $n - 1$. You are probably more familiar with the letters of the orbitals: s, p, d, and f. Answer Question 28 on the Data Sheet.

The shapes of electron orbitals are easier to visualize with the help of animations. Go to this website: *http://www.falstad.com/qmatom/,* and you will see the hydrogenic atom viewer. At the top right, select *"Real Orbitals (chem)."* Then select $n = 1$ and $l = 0$. The next line tells you what orbital these quantum numbers correspond to. Answer Question 29 on the Data Sheet.

Now set $n = 2$ and $l = 1$. Answer Question 30 on the Data Sheet.

Now set $n = 3$ and $l = 2$. Answer Question 31 on the Data Sheet.

Now set $n = 4$ and $l = 3$. Answer Question 32 on the Data Sheet.

# DATA SHEET

Name: _______________________________   Grade: _______________________________

Date Experiment Performed: _______________   Days Late: _______________________

CRN of Lab Section: _____________________   Instructor's Initials: ________________

## Part I: Attractive Potentials

1. If the distance separating the charges in both models is the same, $d_A = d_B$, which has a stronger attractive interaction?

2. If the energy of interaction is the same, $V_A = V_B$, which of the following distance relations is true and why?

   **a.** $d_A = d_B$:

   **b.** $d_A < d_B$:

   **c.** $d_A > d_B$:

## Part II: Basic Concepts of Atomic Structure, Electron Configuration, and Periodicity

3. Which do you expect to be closer to the nucleus, an electron in hydrogen or an electron in helium? Why?

4. The first shell can hold a maximum of two electrons, the second shell can hold eight electrons, and the third can hold eighteen. Draw a shell model diagram for lithium like the ones in Part II for H and He. Don't forget to label the nuclear charge.

   **a.** Li, $Z = 3$

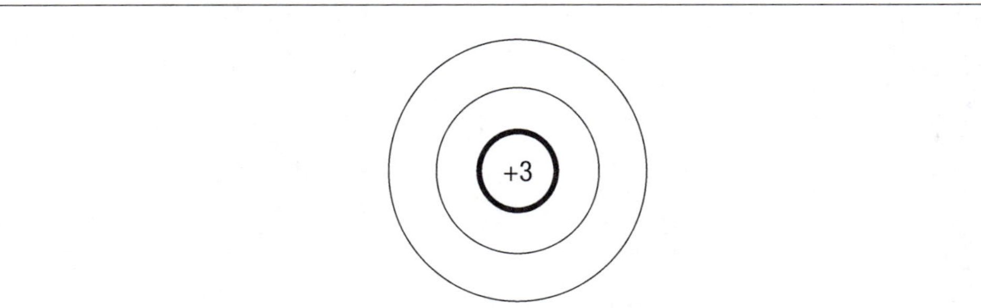


**FIGURE 10.6**

**b.** Ne, $Z$ = 10

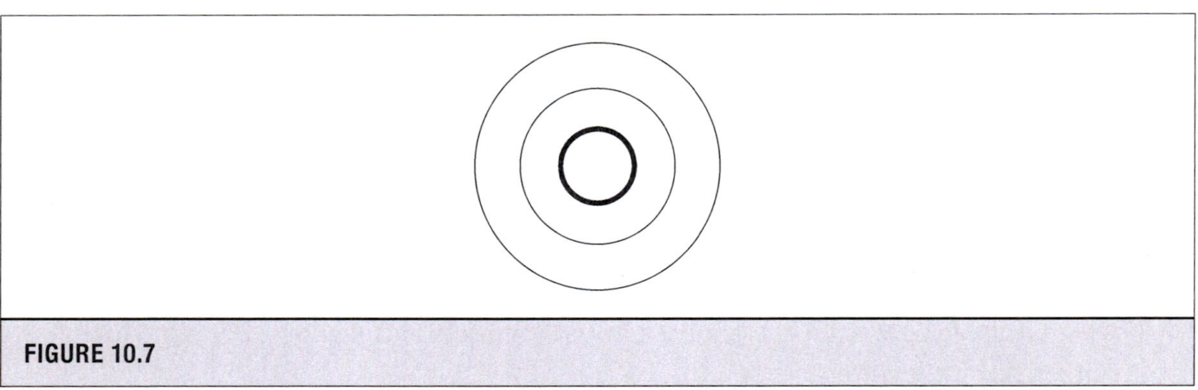

FIGURE 10.7

**c.** Al, $Z$ = 13 (You'll need to draw in a third "shell" below—the first two shells hold 10 electrons total.)

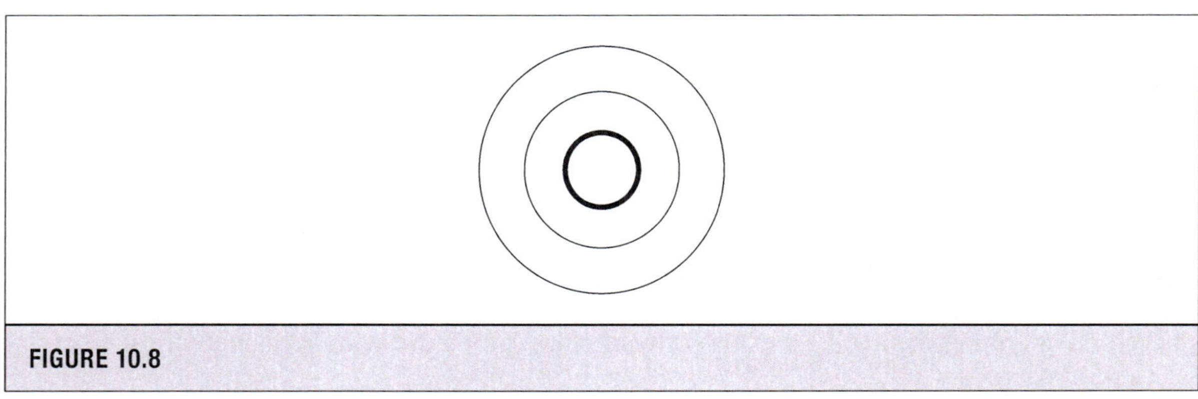

FIGURE 10.8

5. Use Figure 10.9 as you fill in the first six columns of Table 10.2. Do not fill in the last three columns at this time.

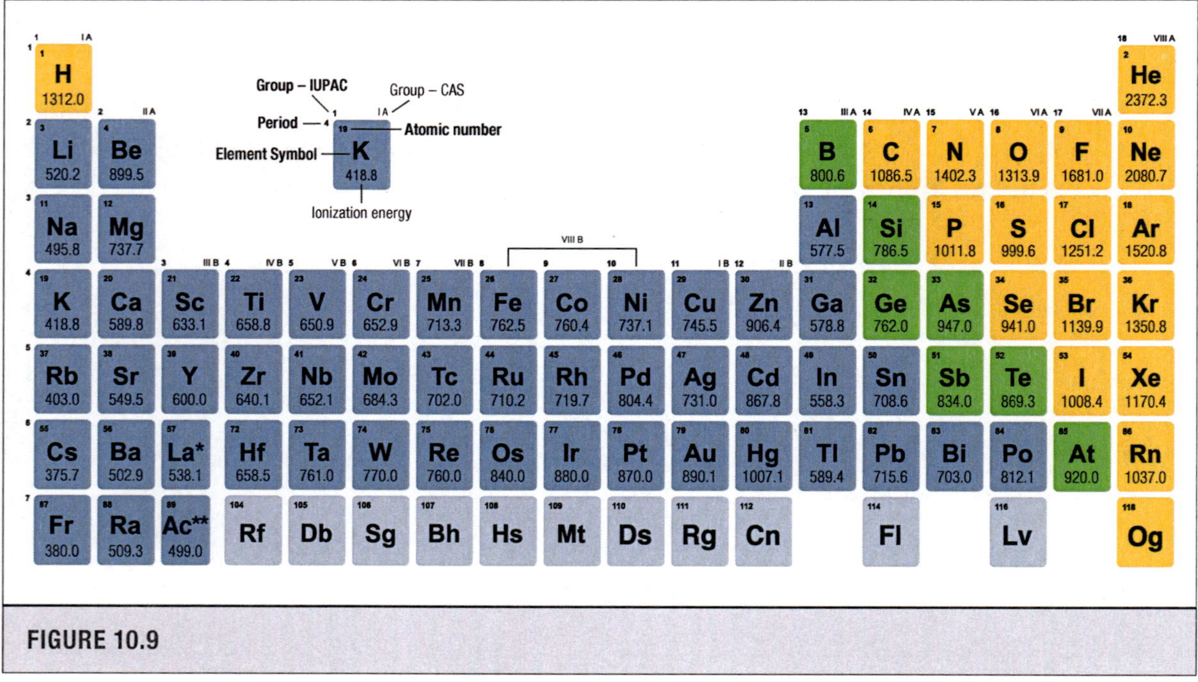



FIGURE 10.9

**TABLE 10.2**

| Element | Atomic Number | # of 1s Electrons | # of 2s Electrons | # of 2p Electrons | # of 3s Electrons | Ionization Energy of Outermost Electron (kJ/mole) | # of Core Electrons | # of Valance Electrons | Core Charge |
|---|---|---|---|---|---|---|---|---|---|
| H | | | | | | | | | |
| He | | | | | | | | | |
| Li | | | | | | | | | |
| Be | | | | | | | | | |
| B | | | | | | | | | |
| C | | | | | | | | | |
| N | | | | | | | | | |
| O | | | | | | | | | |
| F | | | | | | | | | |
| Ne | | | | | | | | | |
| Na | | | | | | | | | |

**6.** What trends do you notice in the ionization energy as the atomic number increases?

## Core Charge

**7.** Use the data in Table 10.2 to answer this question: What is the relationship between the number of valence electrons and core charge of a neutral atom?

**8.** Explain how the core charges of Li and Be are consistent with their ionization energies.

9. Explain how the change in ionization energy from Ne to Na is consistent with the core charge model.

10. The shell number of an atom is designated by the letter $n$. How many electrons does Na have in the shell with $n = 1$? In the shell with $n = 2$? In the shell with $n = 3$?

11. How do the core charges for H, Li, and Na compare to each other? Based on this answer and their respective ionization energies, which species has the valence shell with the largest radius? Which has the valence shell with the smallest radius?

12. Can you deduce a trend in ionization energy as you move from left to right across a period (row)? Can you deduce a trend in core charge as you move from left to right across a period (row)? Explain.

13. Use your textbook (Figure 8.13, page 346) or other resource and locate a table of atomic radii. What is the general trend as you move **from right to left** across a period?

**14.** Explain how the concept of core charge supports this trend. Explain how ionization energy data support this trend.

# Part III: The Electromagnetic Spectrum

**15.** Since electromagnetic radiation has a wave nature, you have to understand the wave concepts of wavelength, frequency, and speed.

   **a.** Label the wavelength of the wave in Figure 10.10.

   **b.** What is the symbol for wavelength?

   **c.** What are the units for wavelength?

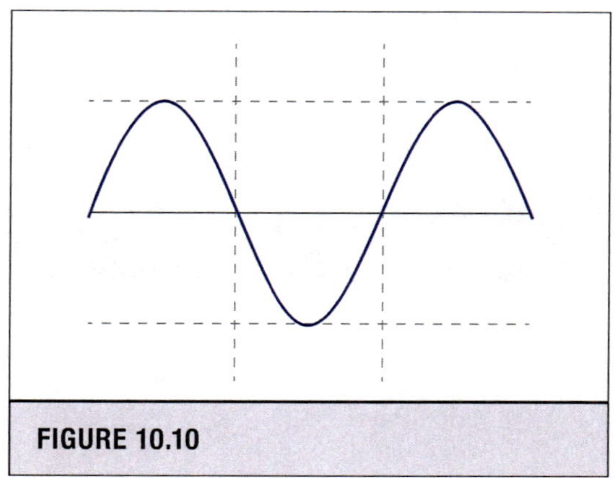

**FIGURE 10.10**

**16.** Define the frequency of a wave.

   **a.** What are the units for frequency?

   **b.** What is the symbol for frequency?

**17.** Define the speed of a wave.

   **a.** What are the units for the speed of electromagnetic radiation?

   **b.** What is the symbol for the speed of electromagnetic radiation?

**18.** Write the equation that relates the wavelength, speed, and frequency of a wave.

**19.** Write the equation that gives the relationship between the energy and the frequency of electromagnetic radiation.

What is the numerical value of $h$, Planck's constant?

## Regions of the Electromagnetic Spectrum

**20.** Use the pentagon shape at the bottom of the figure to slide along the spectrum. In the upper left corner you can see information about the wavelength and frequency of the radiation. Move the pentagon along the spectrum to help you fill in the **blank** spaces in the table below. You will need to calculate the energy column using the equation you gave in Question 19. The first one (gamma rays) has been done for you as an example.

**TABLE 10.3**

| Region | Wavelength (in m) | | Frequency (in Hz) | | Energy (in Joules) | |
|---|---|---|---|---|---|---|
| | Low | High | Low | High | Low | High |
| Gamma Rays (low end only) | $1.55 \times 10^{-10}$ | | $1.94 \times 10^{18}$ | | $1.29 \times 10^{-15}$ | |
| X-Rays | $1.3 \times 10^{-8}$ | | | $1.77 \times 10^{18}$ | | |
| Ultraviolet Radation | | $1.42 \times 10^{-8}$ | | | | |
| Violet Light (high end only) | | | | $1.03 \times 10^{14}$ | | |
| Red Light (low end only) | $1 \times 10^{-5}$ | | | | | |
| Infrared Radiation | | | $3.55 \times 10^{11}$ | | | |
| Mircowave Radiation | $6.53 \times 10^{-1}$ | | | | | |
| Radio Waves (high end only) | | | | $4.21 \times 10^{8}$ | | |

**21. Use the data from Table 10.3 to answer the following questions.**

**a.** Are wavelength and frequency directly related, inversely related, or not related?

**b.** Are energy and frequency directly related, inversely related, or not related?

**c.** Are wavelength and energy directly related, inversely related, or not related?

# Part IV: Electromagnetic Radiation and Atoms

## The Hydrogen Spectrum

**22.** How many electrons does hydrogen have?

**a.** Which energy level is this electron in when the atom is in its ground state?

**23.** On these diagrams, there are three series of lines shown.

**a.** What do all the lines in the Lyman series have in common?

**b.** What do all the lines in the Balmer series have in common?

**c.** What do all the lines in the Paschen series have in common?

**24.** Of these three series of lines, which set has the **highest** transition energies?

**a.** Would this set of lines most likely be in the UV, visible, or IR region?

## Spectra from Other Elements

**25.** Examine the emission spectra for the elements between atomic numbers 1 and 10. (Unfortunately, the link to the emission spectrum for boron is not available.) Which element has the following emission spectrum? Element name:

Red

Orange

Green

Blue

Indigo

Violet

**26.** Which of elements 1–10 has the emission spectrum with the greatest number of lines? How many lines do you see?

# Part V: Quantum Numbers, Orbital Shapes, and Electron Configurations

## The Principal Quantum Number

**27.** The possible values of $n$ range from 1 to 7, corresponding to seven main energy levels—the seven current rows on the periodic chart in our classrooms. On Figure 10.11, label the energy levels with the correct principal quantum number. The nucleus is the solid dot in the center.

## The Angular Momentum Quantum Number

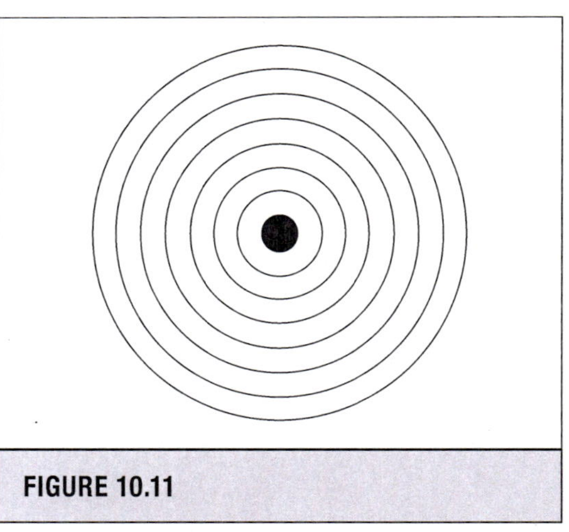

FIGURE 10.11

**28.** Which quantum number goes with each orbital?

s corresponds to an $l$ value of:

p orbitals have an $l$ value of:

d orbtials have an $l$ value of:

f orbitals have an $l$ value of:

**29.** What orbital corresponds to $n = 1$ and $l = 0$? (Give your answer as the principal quantum number and orbital letter. For example, 2p.)

**30.** $n = 2$ and $l = 1$

    **a.** What orbital corresponds to $n = 2$ and $l = 1$?

    **b.** How many orbitals are there for $l = 1$? Use the viewer and look at them.

**31.** $n = 3$ and $l = 2$

    **a.** What orbital corresponds to $n = 3$ and $l = 2$?

    **b.** How many orbitals are there for $l = 2$? Use the viewer and look at them.

**32.** $n = 4$ and $l = 3$

    **a.** What orbital corresponds to $n = 4$ and $l = 3$?

    **b.** How many orbitals are there for $l = 3$? Use the viewer and look.

# Drawing LEDDs of Molecules

# 20-QUESTION PRE-LAB ASSIGNMENT

CRN:________________________________     Your Name: ________________________________

Date Submitted: ________________________     TA's Name: ________________________________

*Please write your answers legibly in the space provided. Many answers will actually consist of several pieces of information that are clearly written on the board during the conversation. You must include all aspects of the answer to receive full credit.* **(You only need to watch this video through the 13:36 time stamp to answer this week's pre-lab activity.)**

**1.** What is the first task listed—the one that you must do this semester?

**2.** What does the abbreviation LEDDs stand for?

**3.** What exactly *is* a Lewis dot structure?

**4.** How do we determine the number of valence electrons?

**5.** Why do we **not** pay attention to the noble gases when drawing LEDDs?

**6.** When we draw in the one dot for the one H electron, could we have placed that dot at any one of the four compass points?

**7.** Why do we typically place that one dot at the "east" position?

**8.** Draw the "traditional" Lewis dot structures for boron, carbon, nitrogen, oxygen, and fluorine.

**9.** When drawing LEDDs for molecules, what is the advantage to using the method that is presented in the video?

**10.** What is the "first" thing that you should do (before you begin drawing a LEDD for a molecule)? Why?

**11.** What is meant by the phrase "arranged in the periphery"?

**12.** Is it *necessary* that the peripheral atoms be arranged "symmetrically"?

**13.** What is the first of the three criteria we must pay attention to? And what is this number for an individual H, N, or C atom? What is the total in both ammonia and methane?

**14.** How many electrons must be shared in the region of space between two atoms?

**15.** What is the second of the three criteria? What do we use to decide how many bonds are expected?

**16.** How many upe's/bonds do we expect for H, N, and C?

**17.** What is the third criterion? What specifically do these rules apply to?

**18.** What is an "octet" in the context of this lab experiment?

**19.** Draw the valid LEDD for methane below. With four covalent bonds to it, how many electrons are around the central carbon atom? How many electrons are associated with each H atom in the LEDD?

**20.** Draw the valid LEDD for ammonia below. With three covalent bonds to it, how many electrons were in place around the central nitrogen atom? What do we have to do in order to get the full octet in place around the nitrogen atom in the LEDD? How many electrons are associated with each H atom in the LEDD?

# PROCEDURE

This handout describes a method that can be used to write valid structural formulas for a majority of molecular formulas.

1. Look at the molecular formula, and decide where to place the atoms in order to draw the LEDD.

    a. The first element listed in the molecular formula will **usually** be placed at the center of the LEDD. (Notable exceptions include compounds with H listed first because, based on valence considerations, H cannot typically be a central atom.)

    b. Elements that there are more than one of in the molecular formula are **rarely** in the center of the LEDD. (Notable exceptions include compounds containing carbon, because carbon atoms have the ability to *catenate.*)

    c. The remainder of the atoms should be symmetrically arranged around the periphery of the central atom.

2. Count up all the **valence** electrons each atom brings into the LEDD. (Recall that the Roman numeral at the top of columns IA–VIIA tells us the number of valence electrons for elements in that family.

3. Begin by placing one **pair** of electrons between the central atom and each peripheral atom.

4. Make sure that each element has its valence requirements satisfied.

5. Check to see that column IVA–VIIA elements satisfy the octet rule and that hydrogen obeys the duet rule.

6. The number of electrons in the final LEDD must equal the total number of valence electrons contributed by all the atoms in the molecular formula.

7. For polyatomic cations, remove one electron per positive charge from the total; for polyatomic anions, add one electron per negative charge to the total.

8. If the octet rule is satisfied, BUT the electron count in the LEDD does not equal the total number of electrons available, then **multiple bonds** will be required in the LEDD.

    1 electron pair = a *single* bond

    2 electron pairs = a *double* bond

    3 electron pairs = a *triple* bond

9. Some column IVA–VIIA elements in Period 3 and below may occasionally exceed the octet rule. (These elements have low-lying d-orbitals which are capable of accommodating the extra electron density.)

10. They may also exceed their expected valence. (Period 2 elements engaged in dative (a.k.a., coordinate covalent) bonding may also exceed their expected valence. But they may *never* exceed the octet!)

11. For any element/atom that *ever* exceeds its expected valence, the actual valence may *never* exceed the actual number of valence electrons that the atom started out with.

12. For CHE 111/112 purposes, the number of valence electrons that should be assigned to transition metals is also the Roman numeral at the top of its column.

13. For CHE 111/112 purposes, the number of "valence electrons" that should be assigned to the noble gases (Kr, Xe) is also the Roman numeral at the top of its column (VIII).

14. The Formal Charge Model can be used to help explain why some atoms violate their expected valence.

15. The sum of the formal charges of all the atoms present must equal the overall charge of the chemical species.

16. The sum of the oxidation numbers of all the atoms present must equal the overall charge of the chemical species.

17. Resonance structures may be necessary to accurately represent empirical evidence like bond angle, bond length, and bond strength.

# DATA SHEET

Name: _________________________________   Grade: _________________________________

Date Experiment Performed: _______________   Days Late: _____________________________

CRN of Lab Section: _____________________   Instructor's Initials: ____________________

| TABLE 11.1   The 14 Molecules and Ions to Be Investigated | | |
|---|---|---|
| **The Central Atom *Does* Obey the Octet Rule** | | |
| $NH_3$ (example) | HBr | $NO_3^-$ |
| $SiH_4$ | CHBrClF | $NH_4^+$ |
| CO | $H_2S$ | $CO_3^{2-}$ |
| **The Central Atom *Does Not* Obey the Octet Rule** | | |
| $ClF_3$ | $PCl_5$ | $AlCl_3$ |
| $I_3^-$ | $SF_6$ | |
| **This Multiple-Centered Molecule *Does* Obey the Octet Rule** (and duet rule for H) | | |
| $CH_3COOH$ | | |

# The Central Atom in Species 1–8
# *Does* Obey the Octet Rule

### 1. Silicon Tetrahydride: $SiH_4$

| TABLE 11.2 | | | | | | |
|---|---|---|---|---|---|---|
| List of Atoms | Number of Atoms | Valence Electrons per Atom | Total Valence Electrons for These Atoms | Ion? (add electrons for negative ion; subtract electrons for positive ion) | Total Valence Electrons (add all of the valence electrons and ion electrons) | Total Number of Bonds and Lone Pairs of Electrons |
| Si | | | | | | |
| H | | | | | | |

**a.** Write an LEDD for each different atom type. (Use the Roman numeral at the top of the column to determine the number of valence electrons.)

What is the electron count? _____________ (Sum up your valence electrons and charges.)

Count the upe's on each atom. That is the predicted valence for each atom. Beneath the LEDD you drew for each atom type above, write the number of expected bonds for each atom.

Identify the atoms that should obey the octet rule (elements in columns IVA–VIIA) _____________, those that may exceed the octet rule (elements in row three and below) _____________, and those that do not satisfy the octet rule (e.g., hydrogen, which satisfies the "duet" rule) _____________.

**b.** Draw a valid Lewis electron dot diagram in the space below.

### 2. Carbon Monoxide: CO

**TABLE 11.3**

| List of Atoms | Number of Atoms | Valence Electrons per Atom | Total Valence Electrons for These Atoms | Ion? (add electrons for negative ion; subtract electrons for positive ion) | Total Valence Electrons (add all of the valence electrons and ion electrons) | Total Number of Bonds and Lone Pairs of Electrons |
|---|---|---|---|---|---|---|
| C | | | | | | |
| O | | | | | | |

**a.** Write an LEDD for each different atom type. (Use the Roman numeral at the top of the column to determine the number of valence electrons.)

What is the electron count? _____________ (Sum up your valence electrons and charges.)

Count the upe's on each atom. That is the predicted valence for each atom. Beneath the LEDD you drew for each atom type above, write the number of expected bonds for each atom.

Identify the atoms that should obey the octet rule (elements in columns IVA–VIIA) _____________, those that may exceed the octet rule (elements in row three and below) _____________, and those that do not satisfy the octet rule (e.g., hydrogen, which satisfies the "duet" rule) _____________.

**b.** Draw a valid Lewis electron dot diagram in the space below.

### 3. Hydrogen Bromide: HBr

**TABLE 11.4**

| List of Atoms | Number of Atoms | Valence Electrons per Atom | Total Valence Electrons for These Atoms | Ion? (add electrons for negative ion; subtract electrons for positive ion) | Total Valence Electrons (add all of the valence electrons and ion electrons) | Total Number of Bonds and Lone Pairs of Electrons |
|---|---|---|---|---|---|---|
| H | | | | | | |
| Br | | | | | | |

**a.** Write an LEDD for each different atom type. (Use the Roman numeral at the top of the column to determine the number of valence electrons.)

What is the electron count? _____________ (Sum up your valence electrons and charges.)

Count the upe's on each atom. That is the predicted valence for each atom. Beneath the LEDD you drew for each atom type above, write the number of expected bonds for each atom.

Identify the atoms that should obey the octet rule (elements in columns IVA–VIIA) _____________, those that may exceed the octet rule (elements in row three and below) _____________, and those that do not satisfy the octet rule (e.g., hydrogen, which satisfies the "duet" rule) _____________.

**b.** Draw a valid Lewis electron dot diagram in the space below.

### 4. Bromochlorofluoromethane: CHBrClF

**TABLE 11.5**

| List of Atoms | Number of Atoms | Valence Electrons per Atom | Total Valence Electrons for These Atoms | Ion? (add electrons for negative ion; subtract electrons for positive ion) | Total Valence Electrons (add all of the valence electrons and ion electrons) | Total Number of Bonds and Lone Pairs of Electrons |
|---|---|---|---|---|---|---|
| C | | | | | | |
| H | | | | | | |
| Br | | | | | | |
| Cl | | | | | | |
| F | | | | | | |

**a.** Write an LEDD for each different atom type. (Use the Roman numeral at the top of the column to determine the number of valence electrons.)

What is the electron count?       (Sum up your valence electrons and charges.)

Count the upe's on each atom. That is the predicted valence for each atom. Beneath the LEDD you drew for each atom type above, write the number of expected bonds for each atom.

Identify the atoms that should obey the octet rule (elements in columns IVA–VIIA)     , those that may exceed the octet rule (elements in row three and below)     , and those that do not satisfy the octet rule (e.g., hydrogen, which satisfies the "duet" rule)     .

**b.** Draw a valid Lewis electron dot diagram in the space below.

### 5. Dihydrogen Sulfide: $H_2S$

**TABLE 11.6**

| List of Atoms | Number of Atoms | Valence Electrons per Atom | Total Valence Electrons for These Atoms | Ion? (add electrons for negative ion; subtract electrons for positive ion) | Total Valence Electrons (add all of the valence electrons and ion electrons) | Total Number of Bonds and Lone Pairs of Electrons |
|---|---|---|---|---|---|---|
| H | | | | | | |
| S | | | | | | |

**a.** Write an LEDD for each different atom type. (Use the Roman numeral at the top of the column to determine the number of valence electrons.)

What is the electron count? ______ (Sum up your valence electrons and charges.)

Count the upe's on each atom. That is the predicted valence for each atom. Beneath the LEDD you drew for each atom type above, write the number of expected bonds for each atom.

Identify the atoms that should obey the octet rule (elements in columns IVA–VIIA) ______, those that may exceed the octet rule (elements in row three and below) ______, and those that do not satisfy the octet rule (e.g., hydrogen, which satisfies the "duet" rule) ______.

**b.** Draw a valid Lewis electron dot diagram in the space below.

## 6. Nitrate Anion: $NO_3^-$

**TABLE 11.7**

| List of Atoms | Number of Atoms | Valence Electrons per Atom | Total Valence Electrons for These Atoms | Ion? (add electrons for negative ion; subtract electrons for positive ion) | Total Valence Electrons (add all of the valence electrons and ion electrons) | Total Number of Bonds and Lone Pairs of Electrons |
|---|---|---|---|---|---|---|
| N | | | | | | |
| O | | | | | | |

**a.** Write an LEDD for each different atom type. (Use the Roman numeral at the top of the column to determine the number of valence electrons.)

What is the electron count? ________ (Sum up your valence electrons and charges.)

Count the upe's on each atom. That is the predicted valence for each atom. Beneath the LEDD you drew for each atom type above, write the number of expected bonds for each atom.

Identify the atoms that should obey the octet rule (elements in columns IVA–VIIA) ________, those that may exceed the octet rule (elements in row three and below) ________, and those that do not satisfy the octet rule (e.g., hydrogen, which satisfies the "duet" rule) ________.

**b.** Draw a valid Lewis electron dot diagram in the space below.

### 7. Ammonium Cation: $NH_4^+$

**TABLE 11.8**

| List of Atoms | Number of Atoms | Valence Electrons per Atom | Total Valence Electrons for These Atoms | Ion? (add electrons for negative ion; subtract electrons for positive ion) | Total Valence Electrons (add all of the valence electrons and ion electrons) | Total Number of Bonds and Lone Pairs of Electrons |
|---|---|---|---|---|---|---|
| N | | | | | | |
| H | | | | | | |

**a.** Write an LEDD for each different atom type. (Use the Roman numeral at the top of the column to determine the number of valence electrons.)

What is the electron count?          (Sum up your valence electrons and charges.)

Count the upe's on each atom. That is the predicted valence for each atom. Beneath the LEDD you drew for each atom type above, write the number of expected bonds for each atom.

Identify the atoms that should obey the octet rule (elements in columns IVA–VIIA)      , those that may exceed the octet rule (elements in row three and below)      , and those that do not satisfy the octet rule (e.g., hydrogen, which satisfies the "duet" rule)      .

**b.** Draw a valid Lewis electron dot diagram in the space below.

## 8. Carbonate Anion: $CO_3^{2-}$

**TABLE 11.9**

| List of Atoms | Number of Atoms | Valence Electrons per Atom | Total Valence Electrons for These Atoms | Ion? (add electrons for negative ion; subtract electrons for positive ion) | Total Valence Electrons (add all of the valence electrons and ion electrons) | Total Number of Bonds and Lone Pairs of Electrons |
|---|---|---|---|---|---|---|
| C | | | | | | |
| O | | | | | | |

a. Write an LEDD for each different atom type. (Use the Roman numeral at the top of the column to determine the number of valence electrons.)

What is the electron count?          (Sum up your valence electrons and charges.)

Count the upe's on each atom. That is the predicted valence for each atom. Beneath the LEDD you drew for each atom type above, write the number of expected bonds for each atom.

Identify the atoms that should obey the octet rule (elements in columns IVA–VIIA)         , those that may exceed the octet rule (elements in row three and below)        , and those that do not satisfy the octet rule (e.g., hydrogen, which satisfies the "duet" rule)        .

b. Draw a valid Lewis electron dot diagram in the space below.

# The Central Atom in Species 9–13
## *Does Not* Obey the Octet Rule

**9. Chlorine TriFluoride: $ClF_3$**

| TABLE 11.10 | | | | | | |
|---|---|---|---|---|---|---|
| List of Atoms | Number of Atoms | Valence Electrons per Atom | Total Valence Electrons for These Atoms | Ion? (add electrons for negative ion; subtract electrons for positive ion) | Total Valence Electrons (add all of the valence electrons and ion electrons) | Total Number of Bonds and Lone Pairs of Electrons |
| Cl | | | | | | |
| F | | | | | | |

**a.** Write an LEDD for each different atom type. (Use the Roman numeral at the top of the column to determine the number of valence electrons.)

What is the electron count?                    (Sum up your valence electrons and charges.)

Count the upe's on each atom. That is the predicted valence for each atom. Beneath the LEDD you drew for each atom type above, write the number of expected bonds for each atom.

Identify the atoms that should obey the octet rule (elements in columns IVA–VIIA)                    , those that may exceed the octet rule (elements in row three and below)                    , and those that do not satisfy the octet rule (e.g., hydrogen, which satisfies the "duet" rule)                    .

**b.** Draw a valid Lewis electron dot diagram in the space below.

### 10. Triiodide: $I_3^-$

**TABLE 11.11**

| List of Atoms | Number of Atoms | Valence Electrons per Atom | Total Valence Electrons for These Atoms | Ion? (add electrons for negative ion; subtract electrons for positive ion) | Total Valence Electrons (add all of the valence electrons and ion electrons) | Total Number of Bonds and Lone Pairs of Electrons |
|---|---|---|---|---|---|---|
| I | | | | | | |

**a.** Write an LEDD for each different atom type. (Use the Roman numeral at the top of the column to determine the number of valence electrons.)

What is the electron count?       (Sum up your valence electrons and charges.)

Count the upe's on each atom. That is the predicted valence for each atom. Beneath the LEDD you drew for each atom type above, write the number of expected bonds for each atom.

Identify the atoms that should obey the octet rule (elements in columns IVA–VIIA)     , those that may exceed the octet rule (elements in row three and below)     , and those that do not satisfy the octet rule (e.g., hydrogen, which satisfies the "duet" rule)     .

**b.** Draw a valid Lewis electron dot diagram in the space below.

## 11. Phosphorous Pentachloride: $PCl_5$

**TABLE 11.12**

| List of Atoms | Number of Atoms | Valence Electrons per Atom | Total Valence Electrons for These Atoms | Ion? (add electrons for negative ion; subtract electrons for positive ion) | Total Valence Electrons (add all of the valence electrons and ion electrons) | Total Number of Bonds and Lone Pairs of Electrons |
|---|---|---|---|---|---|---|
| P | | | | | | |
| Cl | | | | | | |

**a.** Write an LEDD for each different atom type. (Use the Roman numeral at the top of the column to determine the number of valence electrons.)

What is the electron count?       (Sum up your valence electrons and charges.)

Count the upe's on each atom. That is the predicted valence for each atom. Beneath the LEDD you drew for each atom type above, write the number of expected bonds for each atom.

Identify the atoms that should obey the octet rule (elements in columns IVA–VIIA)     , those that may exceed the octet rule (elements in row three and below)     , and those that do not satisfy the octet rule (e.g., hydrogen, which satisfies the "duet" rule)     .

**b.** Draw a valid Lewis electron dot diagram in the space below.

## 12. Sulfur Hexafluoride: $SF_6$

| TABLE 11.13 | | | | | | |
|---|---|---|---|---|---|---|
| List of Atoms | Number of Atoms | Valence Electrons per Atom | Total Valence Electrons for These Atoms | Ion? (add electrons for negative ion; subtract electrons for positive ion) | Total Valence Electrons (add all of the valence electrons and ion electrons) | Total Number of Bonds and Lone Pairs of Electrons |
| S | | | | | | |
| F | | | | | | |

   **a.** Write an LEDD for each different atom type. (Use the Roman numeral at the top of the column to determine the number of valence electrons.)

What is the electron count?           (Sum up your valence electrons and charges.)

Count the upe's on each atom. That is the predicted valence for each atom. Beneath the LEDD you drew for each atom type above, write the number of expected bonds for each atom.

Identify the atoms that should obey the octet rule (elements in columns IVA–VIIA)        , those that may exceed the octet rule (elements in row three and below)       , and those that do not satisfy the octet rule (e.g., hydrogen, which satisfies the "duet" rule)       .

   **b.** Draw a valid Lewis electron dot diagram in the space below.

## 13. Aluminum Chloride: $AlCl_3$

**TABLE 11.14**

| List of Atoms | Number of Atoms | Valence Electrons per Atom | Total Valence Electrons for These Atoms | Ion? (add electrons for negative ion; subtract electrons for positive ion) | Total Valence Electrons (add all of the valence electrons and ion electrons) | Total Number of Bonds and Lone Pairs of Electrons |
|---|---|---|---|---|---|---|
| Al | | | | | | |
| Cl | | | | | | |

**a.** Write an LEDD for each different atom type. (Use the Roman numeral at the top of the column to determine the number of valence electrons.)

What is the electron count? ____________ (Sum up your valence electrons and charges.)

Count the upe's on each atom. That is the predicted valence for each atom. Beneath the LEDD you drew for each atom type above, write the number of expected bonds for each atom.

Identify the atoms that should obey the octet rule (elements in columns IVA–VIIA) ____________, those that may exceed the octet rule (elements in row three and below) ____________, and those that do not satisfy the octet rule (e.g., hydrogen, which satisfies the "duet" rule) ____________.

**b.** Draw a valid Lewis electron dot diagram in the space below.

# Species 14 *Does* Obey the
# Octet Rule (and duet rule for H)

### 14. Acetic Acid: $CH_3COOH$

**TABLE 11.15**

| List of Atoms | Number of Atoms | Valence Electrons per Atom | Total Valence Electrons for These Atoms | Ion? (add electrons for negative ion; subtract electrons for positive ion) | Total Valence Electrons (add all of the valence electrons and ion electrons) | Total Number of Bonds and Lone Pairs of Electrons |
|---|---|---|---|---|---|---|
| C | | | | | | |
| H | | | | | | |
| O | | | | | | |

   **a.** Write an LEDD for each different atom type. (Use the Roman numeral at the top of the column to determine the number of valence electrons.)

What is the electron count?          (Sum up your valence electrons and charges.)

Count the upe's on each atom. That is the predicted valence for each atom. Beneath the LEDD you drew for each atom type above, write the number of expected bonds for each atom.

Identify the atoms that should obey the octet rule (elements in columns IVA–VIIA)      , those that may exceed the octet rule (elements in row three and below)      , and those that do not satisfy the octet rule (e.g., hydrogen, which satisfies the "duet" rule)      .

   **b.** Draw a valid Lewis electron dot diagram in the space below.

# Lab Notebook

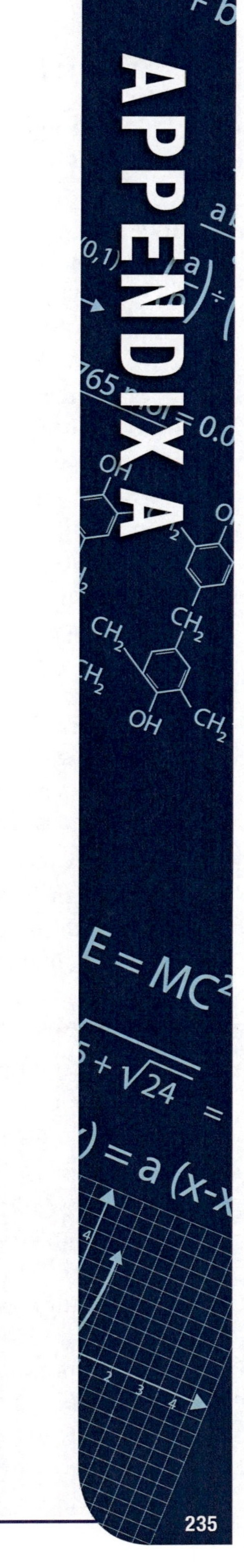

# Important Points about the Lab Notebook

The lab notebook is the official record of the work that a scientist does in the laboratory. The scientist's work is recorded chronologically and documented in ink as it is performed, not after the fact. Well-prepared scientists always have their notebooks at the ready on the spot.

It is therefore prudent to encourage science students very early on to get in the habit of properly preparing and meticulously maintaining a lab notebook. For the vast majority of CHE 111L students, this course represents the very beginning of their collegiate science career. Therefore, CHE 111L (and 112L) students are required to maintain a lab notebook. Your TA and the lab coordinator do not expect the lab notebook to be perfect with the very first attempt. Constructing a lab notebook is a skill and an art that will evolve and mature as you become more experienced.

**One of the most useful ways to keep a lab notebook is to have a space prepared ahead of time in the lab notebook to record every single datum, observation, and measurement that you anticipate you will be collecting during an experiment. The lab report form that will be provided for the majority of the CHE 111L lab experiments is the perfect example of the data you need to be prepared to collect and the calculations you will need to perform. Therefore, the lab report form should initially be emulated, or even copied verbatim, to serve as lab notebook preparation.** As you become more experienced at lab notebook preparation, you will find that it is not always necessary to copy everything over from the lab report form; sometimes it will be necessary to provide space for even more information than the lab report form explicitly requests. But these choices become clearer with experience. **For now, you will never go wrong if you include every item from the lab report form in your lab notebook.**

There is no requirement to copy over a procedure, re-write a procedure, or even to paraphrase the procedure in your lab notebook as long as you can cite the procedure and access it. (On the other hand, if reproducing the procedure in your lab notebook helps you to better understand the experiment you will be doing, there is nothing wrong with writing out the procedure. It is, after all, your lab notebook. However, writing out a procedure in your lab notebook is not required for CHE 111L.)

You must date every page you use and put your name on it. If you work with a partner or two, their name(s) should also be included. The title of the experiment and a purpose must be explicitly stated, followed by the data tables from the lab report form. That is exactly what is required for you to include in your lab notebook. Period.

The lab report form that you submit to your TA at the beginning of the next lab meeting is the culmination of your lab experience. It should, therefore, be legible (in your best penmanship, or hand-printed, or typed if necessary), unblemished (no scratch-outs, no line-outs, nor misspellings), and pristine (on crisp, unwrinkled paper with no "dog ears")—worthy of the importance it represents. Leave your lab report form at home on the day you come to lab. You do not want to risk spilling chemicals on it or accidentally setting it on fire should you inadvertently rest a hot object on it in an effort to protect the lab bench top. Ideally, places to write entries for all the data you need to record on the lab report form is already provided for in your lab notebook.

Safety glasses, prepared lab notebook, copy of the procedure, blue or black ink pen, and calculator—bring these with you to each and every laboratory meeting.

## Sample Experiment and Lab Notebook

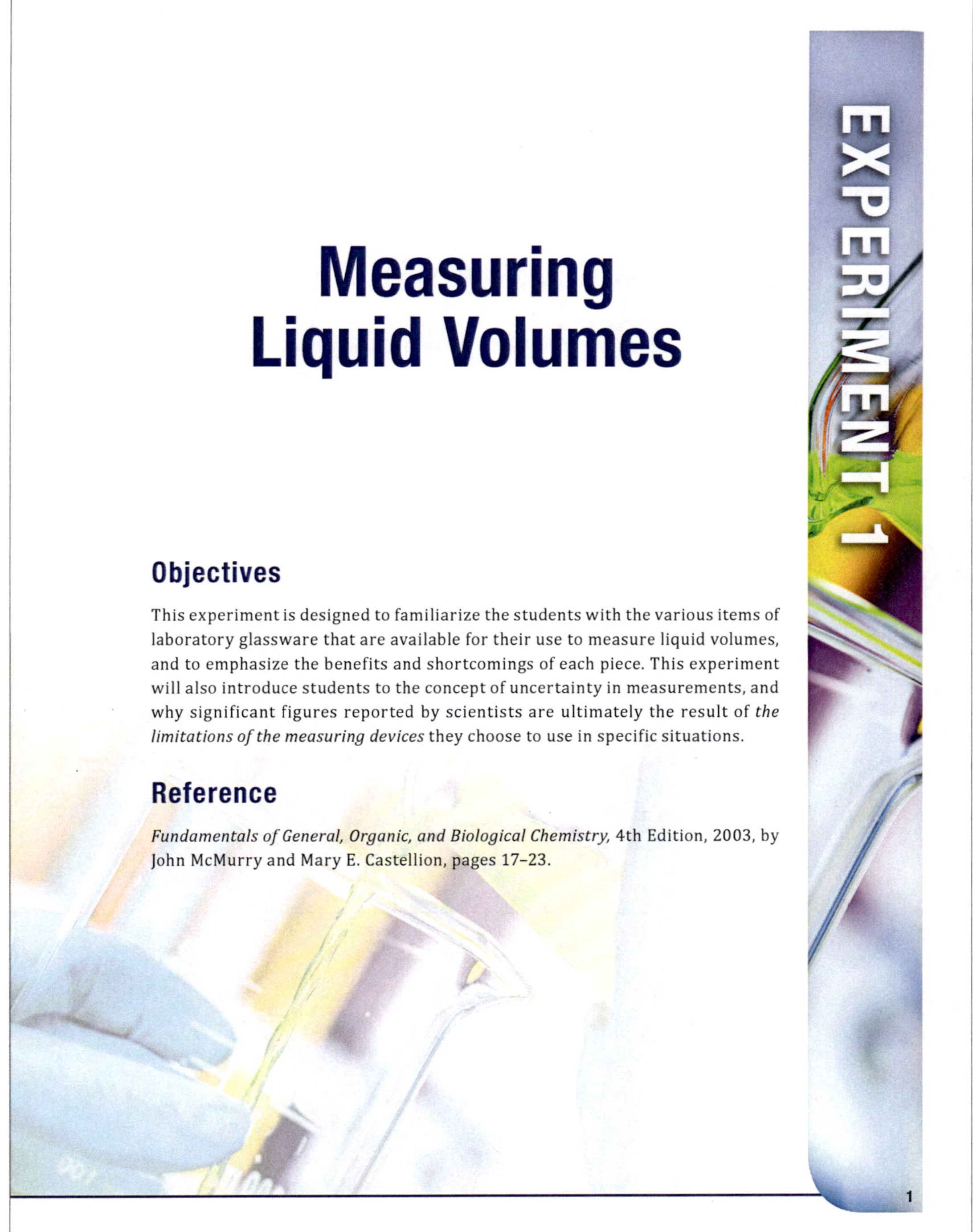


# Measuring Liquid Volumes

### Objectives

This experiment is designed to familiarize the students with the various items of laboratory glassware that are available for their use to measure liquid volumes, and to emphasize the benefits and shortcomings of each piece. This experiment will also introduce students to the concept of uncertainty in measurements, and why significant figures reported by scientists are ultimately the result of *the limitations of the measuring devices* they choose to use in specific situations.

### Reference

*Fundamentals of General, Organic, and Biological Chemistry,* 4th Edition, 2003, by John McMurry and Mary E. Castellion, pages 17–23.

# BACKGROUND

One task that scientist perform repetitively is measuring. Several quantities can be measured in a laboratory, and many different apparati are often available to make the same measurement. You will eventually use the balances in lab to weigh substances such as labware and chemicals. Today you will use beakers, Erlenmeyer flasks, graduated cylinder, burettes, and pipettes to measure liquid volumes. Then we will compare your measurements and discuss how glassware selection is suggested by the sophistication needed for that particular liquid.

2

# PROCEDURE

1. Fill the burette at your station to the 0.00mL graduation with your choice of the colorful liquids. Deliver 50.0 mL of the liquid into a 250-mL beaker. Using the graduations on the beaker, measure and report the volume of the liquid in the beaker.

2. Refill your burette to the 0.00-mL line, and repeat the above process delivering the 50.0 mL of liquid into a 400-mL beaker. Using the graduations on the beaker, measure and report the volume of the liquid in the beaker.

3. Fill a 250-mL beaker to the 100-mL graduation with liquid. Pour this into a tall, 250-mL graduated cylinder at your work bench. Using the graduation on the graduated cylinder, measure and report the volume of the liquid in the cylinder.

4. Repeat the above process, this time filling a 400-mL beaker to the 100-mL graduation with liquid. Then pour the liquid into the tall 250-mL graduated cylinder at your work bench. Using the graduations on the graduated cylinder, measure and report the volume of the liquid in the cylinder.

5. Fill your burette to the 0.00-mL line with liquid. Deliver 20.0 mL of liquid into the 25-mL graduated cylinder. Using the graduations on the graduated cylinder, measure and report the volume of the liquid in the cylinder.

6. Refill your burette to the 0.00-mL line with liquid. Deliver 50.0 mL of liquid into the Erlenmeyer flask at your work bench. Using the graduations on the Erlenmeyer flask, measure and report the volume of the liquid in the flask.

7. Using the 5 mL transfer pipette at your station, transfer exactly 5.00 mL of liquid into your 10-mL graduated cylinder. Using the graduations on the graduated cylinder, measure and report the volume of the liquid in the cylinder.

*Each partner must physically perform each step of the experiment. Be sure to include your own results as well as those of your partner with your lab report.*

# DATA SHEET

Name: ________________________________     Grade: ________________________________

Date Experiment Performed: ________________     Days Late: ________________________

CRN of Lab Section: ________________________     Instructor's Initials: __________________

| TABLE 1.1 | | | | |
|---|---|---|---|---|
| **Delivery Glassware** | **Volume Delivered** | **Receiving Glassware** | **My Volume Measured** | **Partner's Volume** |
| Burette | 50.00 mL | 250-mL Beaker | | |
| Burette | 50.00 mL | 400-mL Beaker | | |
| 250-mL Beaker | 100. mL | 250-mL Graduated Cylinder | | |
| 400-mL Beaker | 100. mL | 250-mL Graduated Cylinder | | |
| Burette | 20.00 mL | 25-mL Graduated Cylinder | | |
| Burette | 50.00 mL | Erlenmeyer Flask | | |
| Pipette | 5.00 mL | 10-mL Graduated Cylinder | | |

| DATE | EXP. NUMBER | EXPERIMENT | | |
|------|-------------|------------|---|---|
| 1/21/14 | I | MEASURING LIQUID VOLUMES | | |
| NAME | | LAB PARTNER | | WITNESS |
| I.M. STUDENT | | M.Y. PARTNER | | |

PURPOSE: To measure liquid volumes using a variety of lab glassware to help me understand the limitations of measuring devices, and to experience first hand the uncertainty that arises due to estimations made when reporting measurements.

PROCEDURE: On the lab handout provided to us for today's lab.

DATA & MEASUREMENTS:

| | DELIVERY | VOL. DEL. | RECEIVING | VOLUME MEASURED | |
|---|----------|-----------|-----------|-----------------|---------|
| | | | | ME | PARTNER |
| 1 | BURETTE | 50.00 | 250-mL BEAKER | | |
| 2 | BURETTE | 50.00 | 400-mL BEAKER | | |
| 3 | 250-mL BEAKER | 100. | 250-mL GRAD. CYL. | | |
| 4 | 400-mL BEAKER | 100. | 250-mL GRAD. CYL. | | |
| 5 | BURETTE | 20.00 | 25-mL GRAD. CYL. | | |
| 6 | BURETTE | 50.00 | E. FLASK | | |
| 7 | PIPETTE | 5.00 | 10 mL GRAD. CYL. | | |